1

Springer Series in Biomaterials Science and Engineering

For further volumes:
http://www.springer.com/series/10955

Series Editor

Prof. Min Wang
Department of Mechanical Engineering
The University of Hong Kong
Pokfulam Road, Hong Kong
e-mail: memwang@hku.hk

Aims and Scope

The Springer Series in Biomaterials Science and Engineering addresses the manufacture, structure and properties, and applications of materials that are in contact with biological systems, temporarily or permanently. It deals with many aspects of modern biomaterials, from basic science to clinical applications, as well as host responses. It covers the whole spectrum of biomaterials – polymers, metals, glasses and ceramics, and composites/hybrids – and includes both biological materials (collagen, polysaccharides, biological apatites, etc.) and synthetic materials. The materials can be in different forms: single crystals, polycrystalline materials, particles, fibers/wires, coatings, non-porous materials, porous scaffolds, etc. New and developing areas of biomaterials, such as nano-biomaterials and diagnostic and therapeutic nanodevices, are also focuses in this series. Advanced analytical techniques that are applicable in R & D and theoretical methods and analyses for biomaterials are also important topics. Frontiers in nanomedicine, regenerative medicine and other rapidly advancing areas calling for great explorations are highly relevant.

The Springer Series in Biomaterials Science and Engineering aims to provide critical reviews of important subjects in the field, publish new discoveries and significant progresses that have been made in both biomaterials development and the advancement of principles, theories and designs, and report cutting-edge research and relevant technologies. The individual volumes in the series are thematic. The goal of each volume is to give readers a comprehensive overview of an area where new knowledge has been gained and insights made. Significant topics in the area are dealt with in good depth and future directions are predicted on the basis of current developments. As a collection, the series provides authoritative works to a wide audience in academia, the research community, and industry.

Iulian Antoniac

Editor

Biologically Responsive Biomaterials for Tissue Engineering

 Springer

Editor
Iulian Antoniac
University Politehnica of Bucharest
Faculty Materials Science and Engineering
Bucharest, Romania

ISBN 978-1-4614-4327-8   ISBN 978-1-4614-4328-5 (eBook)
DOI 10.1007/978-1-4614-4328-5
Springer New York Heidelberg Dordrecht London

Library of Congress Control Number: 2012945004

Printed on acid-free paper

Springer is part of Springer Science+Business Media (www.springer.com)

Preface

The field of biomaterials science represents a unique integration of biology, physics, chemistry, materials science, engineering, and medicine for the purpose of developing new materials solutions to address a variety of medical problems. The past years have witnessed a rapid expansion of the field, and the newest generation of biomaterials, which incorporates biomolecules, therapeutic drugs, and living cells, offers both greater challenges and increased potential for effective medical solutions. Developments in the area of biomaterials, bionanotechnology, tissue engineering, and medical devices are becoming the core of health care. Many critical diseases, such as cancer, cardiovascular disease, and trauma injuries, could be alleviated by regenerative medicine. Almost all medical specialties involve the use of biomaterials, and research plays a key role in the development of new and improved treatment modalities. When a biomaterial is used inside the human body, a biological response is triggered almost instantaneously.

The ultimate goal of tissue engineering is to regenerate tissues that restore normal physiological function, and for this purpose as well as for biocompatibility it is very important for biomaterials to be biologically responsive. An ideal biomaterial for tissue regeneration enhances cell attachment, proliferation, and differentiation, must integrate with the host tissue, and exhibits mechanical properties that are compatible with the native tissue. To reconstruct tissues by tissue engineering, three components are needed: cells, biomaterials, and bioactive molecules. Scaffolds made with different biomaterials play a key role in the organization of new tissue. Also, the manufacturing methods of the scaffold are crucial because the resulting properties need to be correlated with targeted tissue properties.

This book focuses on several current trends in tissue engineering, remodeling, and regeneration. Leading researchers describe the use of nanomaterials to create new functionalities when interfaced with biological molecules or structures. The use of polymers, alloys, and composites, or a combination of these, for biomaterials applications in orthopedics is also explored.

The book is organized into nine chapters which show different aspects of biologically responsive biomaterials for tissue engineering. Key topics include scaffold design, the use of carbon nanotubes in bone cements, endothelial cell response, hydrogel scaffolds processing techniques, synthetic morphogens, bioabsorbable screws, molecular scissors, and bioadhesives.

In addition to coverage of basic science and engineering aspects, a range of applications in bionanotechnology are presented, including diagnostic devices, contrast agents, analytical tools, physical therapy applications, and vehicles for targeted drug delivery.

Biomaterials researchers must be ready to address growing healthcare needs and understand these needs. Innovative biomaterials for tissue engineering will reach the clinic only if these biomaterials will be biologically responsive and if innovative technologies for processing can be translated into industry use. The present volume aspires to bridge the gap between the new biomaterials and processing techniques developed in the laboratory and the clinical effects of currently used biomaterials applications.

These contributions represent essential reading for the biomaterials and biomedical engineering communities and can serve as instructional course lectures targeted at graduate and postgraduate students. This book advocates for biologically responsive biomaterials research, so that emerging technologies can most effectively offer tissue engineering products as clinical solutions for existing diseases. The global imperative for novel biomaterials for tissue engineering is clear, and it is a particularly exciting time to work in the field of biomaterials science for researchers who are successful in achieving this clinical impact.

Bucharest, Romania   Iulian Antoniac

Contents

Contributors

Marco A. Alvarez-Perez Institute of Composite and Biomedical Materials, National Research Council, Naples, Italy

Luigi Ambrosio Institute of Composite and Biomedical Materials, National Research Council, Naples, Italy

Iulian Antoniac University Politehnica of Bucharest, Bucharest, Romania

Thomas Billiet Polymer Chemistry & Biomaterials Research Group, Ghent University, Krijgslaan, Ghent, Belgium

Valentina Cirillo Institute of Composite and Biomedical Materials, National Research Council, Naples, Italy

Cosmin Cotrut University Politehnica of Bucharest, Bucharest, Romania

Carmen Elena Cotrutz Cell Biology Department, G.T. Popa University of Medicine and Pharmacy, Iasi, Romania

Peter Dubruel Polymer Chemistry & Biomaterials Research Group, Ghent University, Ghent, Belgium

Nicholas Dunne School of Mechanical & Aerospace Engineering, Queen's University of Belfast, Belfast, UK

Vincenzo Guarino Institute of Composite and Biomedical Materials, National Research Council, Naples, Italy

C. James Kirkpatrick Institute of Pathology, University Medical Center of the Johannes Gutenberg University, Mainz, Germany

Dan Laptoiu Colentina Clinical Hospital Bucharest, Bucharest, Romania

José Miguel Martín-Martínez Adhesion and Adhesives Laboratory, University of Alicante, Alicante, Spain

Christina A. Mitchell School of Medicine, Dentistry and Biomedical Sciences, Queen's University of Belfast, Belfast, UK

Ross Ormsby School of Mechanical & Aerospace Engineering, Queen's University of Belfast, Belfast, UK

Radu Parpala University Politehnica of Bucharest, Bucharest, Romania

Kirsten Peters Department of Cell Biology, University of Rostock, Rostock, Germany

Tudor Petreus Cell Biology Department, G.T. Popa University of Medicine and Pharmacy, Iasi, Romania

Diana Popescu University Politehnica of Bucharest, Bucharest, Romania

Maria Grazia Raucci Institute of Composite and Biomedical Materials, National Research Council, Naples, Italy

Alfredo Ronca Institute of Composite and Biomedical Materials, National Research Council, Naples, Italy

Matteo Santin Brighton Studies in Tissue-mimicry and Aided Regeneration (BrightSTAR), School of Pharmacy and Biomolecular Sciences, University of Brighton, Brighton, UK

Dieter Scharnweber Max Bergman Center of Biomaterials, Technische Universität Dresden, Dresden, Germany

Jorg Schelfhout Polymer Chemistry & Biomaterials Research Group, Ghent University, Ghent, Belgium

Paul Sirbu Cell Biology Department, G.T. Popa University of Medicine and Pharmacy, Iasi, Romania

Dan Ioan Stoia University Politehnica of Timisoara, Timisoara, Romania

Rafael Torregrosa-Coque Adhesion and Adhesives Laboratory, University of Alicante, Alicante, Spain

Mirela Toth-Tascau University Politehnica of Timisoara, Timisoara, Romania

Roman Tsaryk Institute of Pathology, University Medical Center of the Johannes Gutenberg University, Mainz, Germany

Ronald E. Unger Institute of Pathology, University Medical Center of the Johannes Gutenberg University, Mainz, Germany

Mieke Vandenhaute Polymer Chemistry & Biomaterials Research Group, Ghent University, Ghent, Belgium

Ana María Villarreal-Gómez Bioadhesives Medtech Solutions, Elche, Alicante, Spain

Sandra Van Vlierberghe Polymer Chemistry & Biomaterials Research Group, Ghent University, Ghent, Belgium

Chapter 1
Scaffold Design for Bone Tissue Engineering: From Micrometric to Nanometric Level

Vincenzo Guarino, Maria Grazia Raucci, Marco A. Alvarez-Perez, Valentina Cirillo, Alfredo Ronca, and Luigi Ambrosio

1.1 Introduction

In the natural bone tissue, the ability to sustain characteristic loads is guaranteed either by a tailored hierarchical structure at the micrometer and sub-micrometer scale or by a well-defined chemical composition. In this context, programmed biomechanical properties need to be satisfied since the load bearing function has to be transferred from the engineered material to the growing of the mineralized extracellular matrix (mECM) [13].

In detail, bone tissue is organized into microstructural composites based on the interdigitation of collagen—the most prevalent biopolymer in the body—and an inorganic substance composed of apatitic mineral crystals. The high complexity degree of intrinsic properties of the natural tissues imposes the need to define appropriate strategies for the design of 3-D scaffolds with tailored properties, adapted to be effective in tissue engineering applications. In this context, scaffold design to repair hard tissues such as bone may be complex. Indeed, the primary function of a bone substitute has to be the capability of supporting loads adequately. In that case, the scaffold should not only be a highly porous structure, able to guide novel tissue formation in 3-D space, but also it should have sufficient mechanical strength to offer adequate structural support. Furthermore, it should be able to progressively stimulate new bone growth, by the selective degradation of the polymer phases, resulting, at the end, in a native bone tissue without any trace of the scaffold. As a consequence, scaffolds have to combine several structural and functional properties through an appropriate selection of constituent materials in order to adapt the

V. Guarino (✉) • M.G. Raucci • M.A. Alvarez-Perez • V. Cirillo • A. Ronca • L. Ambrosio
Institute of Composite and Biomedical Materials, National Research Council,
P.le Tecchio 80, Naples 80125, Italy
e-mail: vguarino@unina.it; mariagrazia.raucci@imcb.cnr.it; malvap6@gmail.com;
valentina.cirillo@unina.it; alfredo.ronca@unina.it; ambrosio@unina.it

I. Antoniac (ed.), *Biologically Responsive Biomaterials for Tissue Engineering*,
Springer Series in Biomaterials Science and Engineering 1,
DOI 10.1007/978-1-4614-4328-5_1, © Springer Science+Business Media New York 2013

scaffold features to the requirements of the specific application on micrometric and nanometric scale [15].

Moving towards the current standing of biomaterials science and technology, the challenging way to satisfy the compelling multifunctional needs and extend material property combinations relies on the development of composite materials made of degradable and partially degradable materials by using several technologies for imprinting controlled pore morphology and functional properties [3]. Here, conventional techniques to emboss a controlled pattern of porosity in biodegradable polymers (i.e., phase inversion/salt leaching, rapid prototyping) were preliminarily reviewed, highlighting process–structure–property relationship. Secondly, different strategies to include bioactive surface and bulk signals were proposed in order to improve the osteointegration. Finally, the electrospinning technique was proposed as a powerful technique to develop hierarchically organized nano-structured platforms that closely mimic—biochemically and topologically—the mECM of bone.

1.2   Scaffold Design: Basic Criteria

Scaffold design is becoming essential to the success of scaffold-based tissue engineering strategies. It must combine several structural and functional properties through an appropriate selection of constituent ingredients, namely Elimina virgola cells, materials, and signals, in order to adapt the scaffold features to the requirements of the specific application both on the micrometric and nanometric scale. In particular, an ideal scaffold should possess a repertoire of cues—chemical, biochemical, and biophysical—which are able to control and promote specific events at the cellular (microscale) and macromolecular (nanoscale) level (Fig. 1.1).

Over the last two decades, the concept of cell guidance in tissue regeneration has been extensively discussed. In particular, cell guidance by the scaffold, from cell–material construct to the new engineered tissue, requires a complex balance between chemical, biochemical, and biophysical cues, able to mimic the spatial and temporal microenvironments of the natural extracellular matrix. A relevant part of this evolution concerns the development of novel scaffold materials, compatible with the cell guidance concept and resulting from contemporary advances in the fields of materials science and molecular biology [8].

In particular, the accurate design of materials chemistry allows the attainment of the optimum functional properties that a scaffold can achieve [24]. Materials chemistry contributes to promote cell migration through the scaffold, providing developmental signals to the cells and directing the cell recruitment from the surrounding tissue. However, mass transport requirements for cell nutrition and metabolic waste removal, porous channels for cell migration, and surface characteristics for cell attachment impose the development of tailored porous structures by finely controlled processing techniques. A successful scaffold must also balance architectural features with biological function, allowing a sequential transition in which the regenerated tissue may assume a greater role as the scaffold degrades.

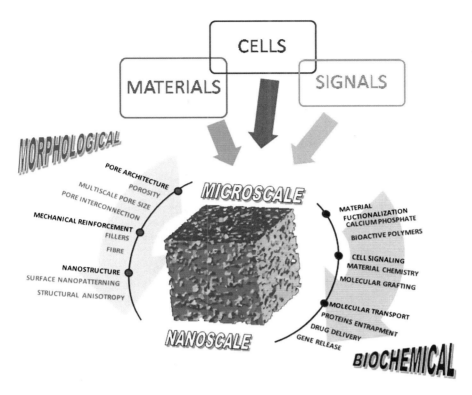

Fig. 1.1 Scheme of basic aspects involved in scaffold design for bone at micrometric and nanometric level

To act as an ECM substitute, a scaffold should impart a 3-D geometry, show appropriate mechanical properties, enable cell attachment, and facilitate the advance of biological events involving cell activities during the formation of a functional tissue. At the microscopic level, a highly porous structure is absolutely essential to support the diffusion of nutrients and waste products through the scaffold. By this time, an optimal pore size is largely recognized in several matrices for cellular response in tissue regeneration (of bone, liver, nerve, cartilage) and for neo-vascularization [44, 55]. Indeed, the optimal pore size tailored on the specific cell type has to be large enough to allow for cell migration and ECM formation yet not be so small that pore occlusion occurs. This balance often presents a trade-off between a denser scaffold which provides better supporting function and a more porous scaffold which enables better interaction with the cell component.

Here, the key issue will be the identification and analysis of the materials and processing techniques which are able to develop micro- and nano-structured platforms. These are necessary to assure an optimal balance in terms of cell recognition, mass transport properties, and mechanical response in order to reproduce the morphological and functional features of new tissue at the microscopic and nanoscopic level.

1.3   Designing Microstructure

To mimic the topological and microstructural characteristics of the ECM, a scaffold has to present high degree of porosity, high surface-to-volume ratio, fully pore interconnection, appropriate pore size, and geometry control [14]. Although several aspects still require further investigations, a large number of studies have just explored the role of the scaffold topographical features in cellular response. These studies demonstrated that the micro-architecture of the porous scaffold may guide cell functions by regulating the interaction between cells, and the diffusion of nutrients and metabolic wastes throughout the three-dimensional (3-D) construct, providing sufficient space for development and later remodeling of the organized tissue [27, 43]. Moreover, morphological features (i.e., pore size) may support, in turn, cell adhesion, molecular transport, vascularization, and osteogenesis. For instance, small pores (with diameters of few microns) favor hypoxic conditions and induce osteochondral formation before osteogenesis occurs [12, 29]. In contrast, scaffold architectures with larger pores (several hundred microns in size) rapidly induce a well-vascularized network and lead to direct osteogenesis [58].

In this context, several technologies are progressively emerging to imprint a controlled pattern of pores within polymeric matrices for triggering the bone formation.

1.3.1   Hierarchically Organized Architecture

Scaffolds for bone regeneration have to provide a highly interconnected porous structure to guide cell ingrowth for tissue formation, while maintaining a sufficient mechanical strength to supply the structural requirement of the substituted tissue. However, the complex composition and structural organization of hard mineralized tissues, such as bone, cannot be replicated using only a single material component which often shows properties spanning in a limited range. In this context, the development of composite scaffolds by combining two or more types of materials with selected properties may represent an appropriate and advantageous solution, satisfying all the mechanical and physiological demands of the host tissue. To date, polymer matrices reinforced by ceramic fillers such as hydroxyapatite and tricalcium phosphates represent a promising material, able to mimic the collagen/hydroxyapatite micro/macro-morphology of "native bone material" [16, 19].

However, the varying success of these approaches may be attributed to a nonhomogeneous spatial distribution of the ceramic particles into the polymer matrix with poor mechanical response and bioactivity. An alternative biomimetic approach is founded on the use of polymer-based composite scaffold with controlled pore spatial distribution and programmed degradation kinetics obtained by integrating polymer continuous fibers. Indeed, continuous fibers integrated within the polymeric matrix are able to mimic the collagen fibers' spatial disposition of the ECM, to guide the preferential cell growth, and to support the stress transfer from the polymer matrix improving the mechanical properties of the scaffold.

Fig. 1.2 Scanning electron microscopy of PCL scaffolds obtained by phase inversion and salt leaching technique

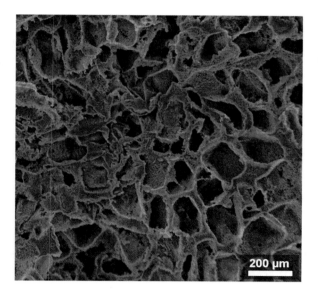

In particular, the synergic combination of phase inversion/salt leaching technique and the filament winding technology allows developing composite scaffolds incorporating polylactide (PLA) continuous fibers within a PCL matrix which satisfy these requirements [17, 18]. The employment of highly biocompatible and bioresorbable PCL and PLA assures the maintenance of sufficient physical and mechanical properties for at least 6 months before their degradation. The integration of a solid porogen (i.e., sodium chloride crystals) within a 3-D polymer matrix enables creation of an interconnected pore network with well-defined pore sizes and shapes (Fig. 1.2).

Tricalcium phosphate (α-TCP) powder, which is able to precipitate in calcium-deficient hydroxyapatite (CDHA) form, is recognized by the host tissue as being similar to natural bone apatite. Moreover, it significantly hinders the stress-shielding phenomena associated with the use of traditional rigid metal-based implants.

Meanwhile, the continuous degradation of the implant causes a gradual load transfer to the healing tissue, preventing the stress-shielding atrophy, with the stimulation of the healing and bone remodeling. The balance between chemical composition and spatial organization of reinforcement systems allows attaining an optimal compromise between mechanical response and bioactive potential to reproduce the bone mECM features.

1.3.2 Multi-scale Degradation

To date, much effort has been dedicated to the design of a variety of composite materials for tissue engineering scaffolds with tailored degradation properties.

Through the time-controlled resizing of pores which occurs during degradation, materials with a predictable degradation rate may guarantee greater penetration of extracellular substance, improve the nutrient/metabolites exchange, and offer a better connection between neighboring cells, encouraging and promoting the growth of new tissue. The influence of several factors on degradation kinetics may be considered, including molecular factors (e.g., chain orientation, molecular weight, and polydispersity), supramolecular factors (crystallinity, spatial distribution of chemically reactive filler), and environmental factors (e.g., mechanical stimuli), to potentially generate a wide range of resorbable properties for custom-made systems. Despite its relevance on affecting specific biological mechanisms at different time scale, the evolution of structural properties in connection with the selective degradation of specific components of the scaffold matrix has not been adequately explored.

The integration of highly degradable materials, obtained by chemical modification of purified hyaluronan (HA), namely, HYAFF11®, formed by the partial or total esterification of the carboxyl groups of glucuronic acid with different types of alcohols gives the opportunity of directly affecting the cell activities, either favoring or, conversely, inhibiting the adhesion of certain cell types [32]. In particular, partially esterified HYAFF11® with reduced hydrophilic, negatively charged, carboxyl groups along the polymer backbone—esterification degree equal to 75 %—shows a peculiar hydrophilic/hydrophobic character which confers an adequate biological recognition, without drastically penalizing the structural integrity of the composite scaffold at the preliminary stage of the culture [21].

In this case, the scaffold functionality is efficiently reached through the right balance between porosity requirements, reinforcement systems, and degradation properties. In particular, the accurate modulation of porosity features and degradation properties on different dimensional scales allows reaching tailored systems able to regulate specific cell mechanisms and to guide the ultimate tissue formation. Also in this case, PLA fibers or bioactive ceramic particles, in turn, further improve the mechanical response of the scaffolds partially limited by the high pore volume fraction, needful for the accommodation of bone cells and their maintenance, matching the mechanical behavior of trabecular bone. Finally, the presence of highly hydrophilic Hyaff11® positively contributes to growth factor entrapment/release as well as to cell recognition, as confirmed by preliminary evaluation of the attachment and MSC cell proliferation, adumbrating the potential in vivo success of tubular scaffolds.

1.4 Surface and Bulk Bioactivation

Novel strategies for regenerating diseased or damaged bone tissue are necessary because of the limitations of established therapies for the treatment of congenital defects and other bone trauma. The development of new materials able to structurally, mechanically, and bio-functionally interface with natural tissues is relevant to the success of regenerative strategies. Synthesis of ceramic/polymer composites represents a fascinating approach for designing "tissue-inspired composite materials" with advantages over either pure ceramic or pure polymer to obtain superior materials for specific applications. Traditionally, calcium phosphate-based ceramics have proved to be attractive materials for biological applications. Among these bio-ceramics, hydroxyapatite (HA)—$Ca_{10}(PO_4)_6(OH)_2$—with an atomic ratio of calcium to phosphorus (Ca/P) of 1.67 has been synthesized by different ways, including aqueous colloidal precipitation and sol-gel methods. They allow synthesizing HA micro/nanoparticles at room temperature, and directly control the particle and grain sizes [42]. In comparison with traditional strategies involving physical mixing of particles, they ensure a more controlled and finer distribution of crystallites in the polymer matrix. They also can enhance the mechanical response in terms of strength, stiffness, toughness, and fatigue resistance to achieve complete mechanical compatibility.

Besides, scaffold design has been also addressed to successfully reproduce the microenvironment required to support and improve the molecular interactions which occur among cells within the mECM [51]. In traditional composite scaffolds—ceramic filler dispersed in a polyester matrix—differences in the chemistry of hydrophilic particles and hydrophobic polymer may lead to inadequate homogenization of the bioactive phase into the polymer matrix, favoring the cluster formation on micrometric scale [42]. Wet chemical precipitation involving the precipitation of HA in tetrahydrofuran, the same solvent utilized to dissolve the polycaprolactone, may be an interesting strategy to overcome these limitations. The use of amphiphilic surfactant (Span85, 75 % oleic acid) aided to improve the dispersion of HA in the PCL matrix. Indeed, oleic acid is a kind of fatty acid with a long hydrocarbon tail and a hydrophilic head [50]. Therefore, oleic acid is believed to mediate the homogenization of hydrophilic HA powders in the pool of a hydrophobic organic solvent, such as tetrahydrofuran [30].

Alternatively, HA/PCL porous scaffolds for bone tissue engineering have been developed by combining chemical synthesis and phase inversion/salt leaching. This strategy allows to reach sub-micrometric HA particles uniformly dispersed within the polymer matrix [46]. Moreover, the use of low temperature approaches to synthesize HA in a solvent in which the polymer is soluble provides a novel pathway to generate homogeneous composite materials preserving the microstructural features of the HA crystals. In this case, the pore architecture was assured by the removal of solvents and used porogens (i.e., NaCl crystals), of predefined particle size. Both chemical and morphological cues are necessary to ensure an adequate biological response following implantation into bone tissue. Indeed, osteoconductivity is essential for the recruitment of cells capable of forming bone matrices and ultimately for successful bone ingrowth into scaffold. Bone invasion in scaffolds can potentially be improved and accelerated by chemical surface modifications, or by the addition of an osteoconductive component in the polymer [56]. To efficiently modify the inner pore wall surfaces, without altering the bulk structure and properties of the scaffolds, a biomimetic approach has been developed based on the surface modification in order to promote the growth of apatite-like crystals onto polymeric pore surface by Simulated Body Fluid (SBF) treatment [58] with an ions concentration, at least, equal to that of the human blood plasma [1]. However, some authors have proposed to accelerate the conventional biomimetic process from 7 to 28 days down to few days by using supersaturated SBF, thus confirming the interdependence between ionic strength and pH, as well as the influence of the substrate chemistry on the formation of the resultant apatite [4].

Alternatively, some authors proposed to combine bulk sol-gel transition and surface biomimetic techniques. The sol-gel reaction allows to finely disperse calcium phosphate nanoparticles into a polycaprolactone (PCL) matrix with an expected improving of the functional features (i.e., mechanical response) as reported in previous works [35]. Meanwhile, the controlled deposition of HA crystals by accelerating simulated body fluid (5× SBF) allows forming a biologically active surface which can drastically improve the bonding to living bone [23]. Proposed integrated approach enables to efficiently exploit the peculiar features of composite material bulk and surfaces by modulating the spatial distribution of bioactive signals, moving towards a more efficient replacement of bone tissue.

Since the apatite particles are generated from aqueous solution, this technique may be used for biomimetically coating porous scaffolds with highly complex surfaces, unlike many other surface modification methods which are restricted to flat surfaces or very thin porous layers. This technique involves the use of biocompatible solution at mild conditions (body temperature reaction), extending the variety of polymers or composite materials to be potentially used. Moreover, apatite crystals are more similar to natural bone mineral in terms of low crystallinity and nanometric size of crystals, beneficially affecting degradation and remodeling properties. *In vitro* experiments showed that mineral crystals grow on composite materials just after 7 days in supersaturated biological solution (5× SBF) [47]. It was not surprising that higher concentration facilitated apatite formation, since it was reported [58] that increasing the ionic concentrations of an SBF above the saturation limit is a practical way to facilitate apatite nucleation and growth.

Other recent studies have reported that the formation of apatite on artificial materials is induced by functional groups which could reveal negative charge and further induce apatite via the formation of amorphous calcium phosphate (Fig. 1.3). After 7 days of soaking in 5× SBF, the composite material was uniformly covered with a calcium phosphate crystals layer with a thickness of few microns. In particular, rose petal-like apatite crystallites composed mainly of hydroxyapatite are detected with a Ca/P molar ratio of 1.67, typical of natural hydroxyapatite. Moreover, in this case, a large presence of HA crystals has been recognized along the inner pore walls of the porous scaffolds, confirming that the high concentration of SBF solution is capable to induce a faster and uniformly distributed deposition. In particular, the copious formation of apatite-like globular crystals on the exposed surfaces of the composite scaffolds will be directly correlated to the improvement offered by nanoscaled HA particles on the mechanisms of osteointegration [47].

However, long treatment times (7–14 days) often are not completely compatible for tissue engineering scaffolds which are composed of highly degradable

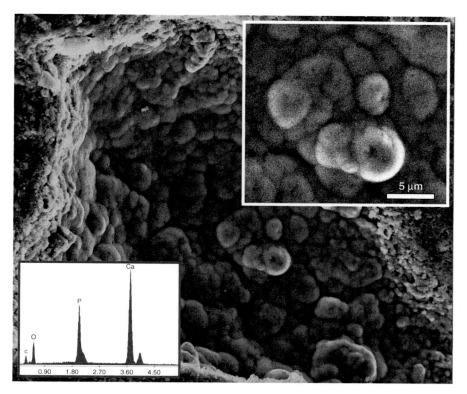

Fig. 1.3 Bioactive scaffolds obtained by surface mineralization via SBF treatment: cross section images of apatite crystal coating onto the pore surface, particularly of globular apatite crystals (*upper square*) and EDS spectra for Ca and P detection (*bottom square*)

polymers like Hyaluronic Acid derivate (HYAFF11®). Indeed, HYAFF11® dissolves in water solution only after few weeks of treatment, showing a partial de-esterification in the presence of basic pH. In particular, Campoccia et al. [6] have demonstrated that benzyl ester HYAFF11® underwent a spontaneous hydrolytic degradation of the ester bonds even in the absence of any enzymatic activity with a complete dissolution just after 1 week. This explains the relevant effect of esterification degree on the degradation mechanisms as well as on the hydrophilic behavior of the final material. Indeed, as the esterification degree increases, polymer chains become progressively less flexible and hydrated, but also more rigid and stable, limiting the hydrolytic attacks which act as more active nucleation initiation sites than microscale particles. In this case, Kokubo-like strategies rapidly fail due to a lack of crystal deposition in used experimental window time. It means that a drastic reduction of exposure time is mandatory to prevent a catastrophic uncontrolled material loss during the conditioning via SBF solution. Some authors have also demonstrated that organic polymers can also form a bone-like apatite layer on their surface in the presence of highly supersaturated solutions with ion concentration from one- to fivefold SBF salt concentration [4]. Furthermore, several studies have demonstrated that an effective apatite growth can occur in the presence of specific environmental conditions which influence the chemical solubility as well as the material surface activation [40].

In this case, a novel treatment combining the preliminary use of bioglasses with supersaturated SBF solutions at different salt concentrations has been optimized to overcome the applicability limitations of traditional treatments on highly degradable polymers like HYAFF11®. The analysis performed on this biomimetic HA-based scaffold evidenced that the proposed treatment supports the mineralization occurring already at day 21 and significantly increased at day 35 [39]. Apatite crystals on HYAFF11® scaffold seem to act as nucleation sites for the deposition of inorganic bone components as we proved by light and electron microscopy. In fact, on non-biomimetic HYAFF11® scaffold an immature mineralization started to occur only at day 35, where we have also found a lower Collagen type I expression and increased cell proliferation, indicating that the biomimetic treatment reduced the time taken to induce the process. Contrariwise, biomimetic HYAFF11® scaffold promotes the precipitation of apatite crystals, inducing the formation of new bone with a copious mineralization already at day 21 increasing up to day 35. These data suggest possible in vitro trophic effects of h-MSCs [7, 10] in inducing mineralization processes that can be also influenced by the composition [22] or elasticity [48] of the scaffold. In fact, HA not only enhances hydroxyapatite crystal proliferation and growth, but it can also promote mineralization [5] since it is a prominent extracellular matrix component during the early stage of osteogenesis [53].

1.5   Designing Nanostructures

In biological and medical applications, the capability of controlling physical and chemical interactions at the level of natural building blocks, from proteins to cells, may favor a more efficient exploration, manipulation, and application of living systems and biological phenomena. In this context, nano-structured biomaterials in the form of nanoparticles, nanofibers, nanosurfaces, and nanocomposites have gained increasing interest in regenerative medicine because they often mimic the physical features of natural ECM at the nanoscale level.

Besides, nanotechnology is emerging as an important tool in scaffold design to reproduce the features of microenvironment-mediated signaling which determine tissue specificity and architecture of native tissues. Currently, an important class of nano-structured biomaterials on which intensive research has been carried out is composed of nanofibrous materials, especially biodegradable polymer nanofibers, able to morphologically mimic the fibrillar structure of ECM. In this regard, the electrospinning represents a powerful strategy to develop biomimetic fibrous material to be efficaciously used as scaffold for the ECM. Electrospinning is a simple and versatile technique for producing nonwoven, interconnected nanofiber mats that have potential applications in the field of engineering and medicine. Traditionally, electrospinning has been used to generate platforms on submicron or nanometric scale in order to overcome the limitation connected to traditional fiber drawing processes. Today, the current surge in nanotechnology has adopted electrospinning as an elegant technique to produce nanofibrous structures for biomedical applications such as wound healing, tissue engineering, and drug delivery. In comparison with conventional techniques, electrospinning has demonstrated its ability to fabricate nanofiber scaffolds that closely mimic the native ECM structure [26]. This technique exhibits a large versatility to produce scaffolds with a range of mechanical properties (either in plastic and elastic field), optimized porosity, and pore volume for tissue engineering applications [34, 41]. Here, different polymer and composite electrospun membranes with various micro- or nano-metric architecture were investigated in order to analyze the contribution of fiber patterning and biochemical signals to the biological response of mesenchymal stem cells (MSCs).

1.5.1   Micro or Nanotexture?

In the biomedical field, micro- or nanoscale textured surfaces have gained important attention, because the majority of tissues present in ECM commonly composed of collagen fibers hierarchically assembled into dense bundles. The mimicking of the structural organization firstly is due to the spatial distribution of collagen fibers into the extracellular matrix. The need of regenerating these structurally complex tissues imposes to develop novel technologies which can impart multi-scale organization of fibers from the micron and nanometric level. Electrospinning is a versatile technique which promises to produce fibers on the micro or nano-length scale with tailoring physical, chemical, and biological properties, largely adaptable to various cellular environments for specific biomedical applications.

The ability of modulating the fiber size to reach the desired morphology depends upon a large number of variables including materials (i.e., polymer concentration, solvent chemistry) and process parameters (i.e., voltage, flow rate). In particular, the creation of a microstructured fibers network based on synthetic polymers (i.e., PCL) provides to support mechanical properties while nanotexture firstly contributes to cell–material interaction by offering a higher surface for binding sites to cell membrane receptors promoting cellular attachment, proliferation, and growth.

From a biological point of view, it is generally recognized the relevance of micro- and nano-patterning and texture size scale on the cell–materials interaction mechanisms may be crucial to improve the cell adhesion and spreading [11]. Several studies also demonstrated that the variation of the characteristic size scale of fiber meshes obtained by an accurate control of fiber diameter significantly influences the cytocompatibility [28]. In particular, it has been proven that nanotopography provides several effects on the cell shape, morphology, adhesion, and proliferation. Osteoblasts adhesion was improved when they were cultured on nanotexture surfaces compared to the conventional micro surfaces [57]. NIH-3T3 cells cultured onto PGA/collagen fibers with submicrometric size scale appear fully attached, elongated, and more fibroblastic than those seeded on micrometric fibers [52]. SaOs-2 and BMSC showed an influence on cytoskeletal organization and cell viability when cultured onto nanofibrous scaffold [54]. All these observations from current literature have been also confirmed by the comparative studies of casted films and PCL electrospun fibers with micro- and nano-texture [21]. Preliminary, it has been shown that cell adhesion and viability of human MSCs are significantly enhanced by the nanometric scale of fibers which offer a more efficacious anchorage to cells. Moreover, nanotopography also improves proliferation and differentiation efficiency of MSCs, thus offering an active topographical signal able to influence cell differentiation. In particular, it has been reported that fibers nanotexture is also capable of supporting differentiation of MSCs into osteogenic phenotype, consistently with others' reported literature. These concur to demonstrate the relevance of nanotopography on guiding the fate of the osteogenic, chondrogenic, and neurogenic lineages of MSCs [25, 45].

1.5.2 Chemical Cues for Cell Recognition

Cell–substrate interaction may be drastically affected by the presence of chemical cues able to support all the main cell functions including adhesion, proliferation, and differentiation of cells. An interesting approach for bone tissue engineering certainly relies on the ability to geometrically and topologically reproduce the native microenvironment of tissues, providing all the biochemical stimuli which characterize the extracellular matrix of natural tissues. In order to afford the effective biomimesis of the natural ECM both from morphological and biochemical point of view, a new strategy to designing micro- and nano-fibers polymers scaffolds involves the integration of natural biopolymers able to enhance the recognition of the fiber surface due to their innate hydrophilic nature.

Intrigued by their high biological recognition, natural polymers such as collagen, gelatine, elastin, silk fibroin, fibrinogen, hyaluronan, and chitosan have been just fabricated into three-dimensional nanofibrous scaffolds for tissue engineering applications. Based on this approach, various polymer compositions such as synthetic/natural polymer blends were recently electrospun to produce new scaffolds with required bioactivity and mechanical properties suitable for vascular, dermal, neural, and cartilage tissue engineering [9, 33].

For instance, PCL may be successfully electrospun with gelatin from a (50/50 wt/wt) polymer solution in fluorinated solvents (i.e., TFE, HFP) for obtaining a fiber network with submicrometric sizes and random fiber distribution (Fig. 1.4).

In this case, nanofibers exhibited good biocompatibility of MSCs in terms of adhesion, spreading, and proliferation. Although synthetic polymers like PCL are

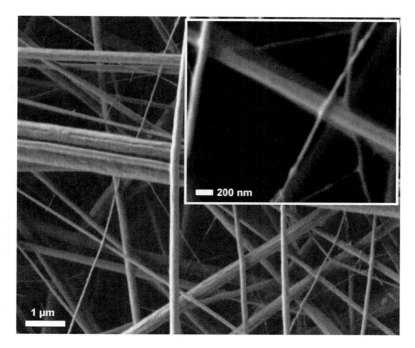

Fig. 1.4 PCL/gelatin nanofibers by electrospinning: SEM images at low and high (*square*) magnification

intrinsically able to promote cell adhesion and direct the growth of cells, cell affinity is partially compromised by their low hydrophilicity and lack of surface cell recognition. Thus, the integration of a gelatin phase significantly improves the hydrophilic property, also supporting the more common strategies of encoding cell-recognition domains [2, 20]. The presence of gelatin molecules improves the cell affinity with the hydrophobic PCL due to the exposure of preferential integrin sites able to favor cell adhesion and differentiation, thus creating favorable environments which reproduce the chemical cues of natural extracellular matrix. These results are in agreement with several studies on the nanofiber blends obtained by endowing gelatin into synthetic polymers. For instance, PCL/gelatin nanofibers promote BMSC growth and migration [59]. Bone cells both cultured onto PLA/gelatin and PLA fibers clearly indicate a higher viability of osteoblasts (MC3T3-E1) in the presence of protein cues [31]. Furthermore, PCL nanofibers with heparin sulfate with MSCs pre-committed to an osteogenic lineage promote the secretion of bone matrix and the formation of newly bone tissue under a subcutaneous model in nude mice [36].

Besides, different approaches based on the mixing of natural and synthetic polymers have been also investigated with successful results. For instance, synthetic nanofibers coated with natural polymers, such as collagen and gelatin, offer an active surface able to promote a good initial adhesion and growth of osteoblasts [37, 38, 49]. Moreover, the further integration of calcium carbonate, calcium phosphate, and hydroxylapatite clays into nanofibrous polymer matrix allows mimicking the natural composition of the bone ECM. In particular, nano-HA particles also concur to slow down the degradation process by neutralizing or buffering the pH changes caused by the typical acidic degradation products of polyesters, potentially enhancing mechanical and biological response of the scaffolds during in vitro and in vivo implant. Synthetic polymers endowed to nano-sized HA particles offer a chemical cue able to affect the cell behavior, thus catalyzing bioactivity and mineralization processes. Although the underlying mechanism for this phenomenon is not clear, bioactive HA particles obtained by the dissolution and precipitation of calcium and phosphate ions out of the fiber surface layer provide to create a microenvironment that resembles the natural bone and stimulates osteogenic differentiation of MSCs.

1.6   Conclusions

Micro- or nanostructured tissue engineering scaffolds have the potential to direct cell destiny, as well as to regulate processes such as angiogenesis and cell migration.

Architecture and chemical cues concur to induce cells' attachment and trigger specific cell/materials interaction mechanisms. The addition of polymer phases may positively contribute to finely control morphology, mechanical response, degradation profile, and, as a consequence, cellular events, guiding new tissue ingrowth by the gradual transmission of biochemical and biophysical signals, moving towards a true mimicking of natural ECM functions.

Acknowledgments This study was supported by the Italian Research Network "TISSUENET" n. RBPR05RSM2 and by IP STEPS EC FP6-500465.

References

1. Abe Y et al (1990) Apatite coating on ceramics metals and polymers utilizing a biological process. J Mater Sci Mater Med 1(4):233–238
2. Alvarez-Perez MA et al (2010) Influence of gelatin cues in PCL electrospun membranes on nerve outgrowth. Biomacromolecules 11(9):2238–2246
3. Ambrosio L et al (2001) In: Chiellini et al (eds) Biomedical polymers and polymer therapeutics. Kluwer Academic/Plenum Publishers, US Springer Part 1, pp 227–233
4. Barrere F et al (2002) Nucleation of biomimetic Ca-P coatings on Ti6Al4V from a SBFx5 solution: influence of magnesium. Biomaterials 23(10):2211–2220
5. Boskey AL et al (1991) Hyaluronan interactions with hydroxyapatite do not alter in vitro hydroxyapatite crystal proliferation and growth. Matrix 11:442–446
6. Campoccia D et al (1998) Semi-synthetic resorbable materials from hyaluronan esterification. Biomaterials 19:2101–2127
7. Caplan AI et al (2006) Mesenchymal stem cells as trophic mediators. J Cell Biochem 98:1076–1084
8. Causa F et al (2007) A multi-functional scaffold for tissue regeneration: the need to engineer a tissue analogue. Biomaterials 28:5093–5099

9. Chong EJ et al (2007) Evaluation of electrospun PCL/gelatin nanofibrous scaffold for wound healing and layered dermal reconstitution. Acta Biomater 3:321–330
10. Djouad F et al (2009) Mesenchymal stem cells: innovative therapeutic tools for rheumatic diseases. Nat Rev Rheumatol 5:392–399
11. Garcia AJ et al (1999) Integrin-fibronectin interactions at cell-material interface: initial integrin binding and signaling. Biomaterials 20(23–24):2427–2433
12. Griffith LG et al (2002) Tissue engineering: current challenges and expanding opportunities. Science 295:1009–1014
13. Guarino V et al (2007) Bioactive scaffolds for bone and ligament tissue. Exp Rev Med Devices 4(3):405–418
14. Guarino V et al (2007) Porosity and mechanical properties relationship in PCL based scaffolds. J Appl Biomater Biomech 5(3):149–157
15. Guarino V et al (2008) Design and manufacture of microporous polymeric materials with hierarchal complex structure for biomedical application. Mater Sci Tech 24(9):1111–1117
16. Guarino V et al (2008) The role of hydroxyapatite as solid signal on performance of PCL porous scaffolds for bone tissue regeneration. J Biomed Mater Res B Appl Biomater 86B:548–557
17. Guarino V et al (2008) Polylactic acid fibre reinforced polycaprolactone scaffolds for bone tissue engineering. Biomaterials 29:3662–3670
18. Guarino V et al (2008) The synergic effect of polylactide fiber and calcium phosphate particles reinforcement in poly ε-caprolactone based composite scaffolds. Acta Biomater 4(6):1778–1787
19. Guarino V et al (2009) The influence of hydroxyapatite particles on "in vitro" degradation behaviour of PCL based composite scaffolds. Tissue Eng Part A 15(11):3655–3668
20. Guarino V et al (2009) Polycaprolactone and gelatin electrospun platforms for bone regeneration. Reg Med 4(6 Suppl 2):S187–S188
21. Guarino V et al (2010) Morphology and degradation properties of PCL/HYAFF11-based composite scaffolds with multiscale degradation rate. Comp Sci Tech 70:1826–1837
22. He J et al (2010) Osteogenesis and trophic factor secretion are influenced by the composition of hydroxyapatite/poly(lactide-co-glycolide) composite scaffolds. Tissue Eng Part A 16:127–137
23. Hench LL (1991) Bioceramics: from concept to clinic. J Am Ceram Soc 74(7):487–510
24. Hollister SJ (2005) Porous scaffold design for tissue engineering. Nat Mater 4:518–524
25. Hu J et al (2009) Chondrogenic and osteogenic differentiations of human bone marrow-derived mesenchymal stem cells on nanofibrous scaffold with designed pore network. Biomaterials 30:5061–5067
26. Huang Z et al (2003) A review on polymer nanofibers by electrospinning and their applications in nanocomposites. Compos Sci Technol 63:2223–2253
27. Hutmacher DW et al (2007) State of the art and future directions of scaffold-based bone engineering from a biomaterials perspective. J Tissue Eng Regen Med 1:245–260
28. Jager M et al (2007) Significance of nano and microtopography for cell surface interactions in orthopaedic implants. J Biomed Biotech 2007. doi:101155/2007/69036
29. Karande TS et al (2004) Diffusion in musculoskeletal tissue engineering scaffolds: design issues related to porosity permeability architecture and nutrient mixing. Ann Biomed Eng 32:1728–1743
30. Kim HW et al (2007) Nanofibrous matrices of poly(lactic acid) and gelatin polymeric blends for the improvement of cellular responses. J Biomed Mater Res A 87A:25–32
31. Kim HW (2007) Biomedical nanocomposites of hydroxyapatite/polycaprolactone obtained by surfactant mediation. J Biomed Mater Res 83A:169–177
32. Kon E et al (2009) Tissue engineering for total meniscal substitution: animal study in sheep model. Tissue Eng Part A 14(6):1067
33. Li M et al (2006) Electrospinning polyaniline contained gelatin nanofibers for tissue engineering applications. Biomaterials 27:2705–2715
34. Li WJ, Laurencin CT, Caterson EJ, Tuan RS, Ko FK (2002) Electrospun nanofibrous structure: a novel scaffold for tissue engineering. J Biomed Mater Res 60:613–621
35. Liu Q et al (1997) Nano-apatite/polymer composites: mechanical and physiochemical characteristics. Biomaterials 18(19):1263–1270

36. Luong-Van E et al (2007) The in vivo assessment of a novel scaffold containing heparan sulfate for tissue engineering with human mesenchymal stem cells. J Mol Histol 38:459–468
37. Ma K et al (2005) Grafting of gelatin on electrospun poly(caprolactone) nanofibers to improve endothelial cell spreading and proliferation and to control cell orientation. Tissue Eng 11:1149–1158
38. Ma K et al (2008) Electrospun nanofiber scaffolds for rapid and rich capture of bone marrow-derived hematopoietic stem cells. Biomaterials 29:2096–2103
39. Manferdini C et al (2010) Mineralization behavior with mesenchymal stromal cells in a bio-mimetic hyaluronic acid-based scaffold. Biomaterials 31(14):3986–3996
40. Mao C et al (1998) Biomimetic growth of calcium phosphates with an organized hydroxylated surface as template. J Mater Sci Lett 17:1479–1481
41. McManus MC et al (2006) Mechanical properties of electrospun fibrinogen structures. Acta Biomater 2:19–28
42. Mobasherpour I et al (2007) Synthesis of nanocrystalline hydroxyapaptite by using precipita-tion method. J Alloys Compd 430:330–333
43. Ng AMH et al (2008) Differential osteogenic activity of osteoprogenitor cells on HA and TCP/HA scaffold of tissue engineered bone. J Biomed Mater Res 85A:301–312
44. O'Brien FJ et al (2005) The effect of pore size on cell adhesion in collagen-GAG scaffolds. Biomaterials 26:433–441
45. Prabhakaran MP et al (2009) Mesenchymal stem cells differentiation to neuronal cells on electrospun nanofibrous substrates for nerve tissue engineering. Biomaterials 26:2603–2610
46. Raucci MG et al (2010) Biomineralized porous composite scaffolds prepared by chemical synthesis for bone tissue regeneration. Acta Biomater. doi:101016/jactbio201004018
47. Raucci MG et al (2010) Hybrid composite scaffolds prepared by sol-gel method for bone regeneration. Comp Sci Technol. doi:101016/jcompscitech201005030
48. Seib FP et al (2009) Matrix elasticity regulates the secretory profile of human bone marrow-derived multipotent mesenchymal stromal cells (MSCs). Biochem Biophys Res Commun 389:663–667
49. Shin YRV et al (2006) Growth of mesenchymal stem cells on electrospun type I collagen nanofibers. Stem Cells 24:2391–2397
50. Small DM (1986) The physical chemistry of lipids: from alkanes to phospholipids. Plenum, New York
51. Stevens MM et al (2005) Exploring and engineering the cell interface. Science 310:1135–1138
52. Tian F et al (2008) Quantitative analysis of cell adhesion on aligned micro- and nanofibers. J Biomed Mater Res 84A:291–299
53. Toole BP et al (1972) Hyaluronate in morphogenesis: inhibition of chondrogenesis in vitro. Proc Natl Acad Sci U S A 69:1384–1386
54. Tuzlakoglu K et al (2005) Nano and micro-fiber combined scaffold: a new architecture for bone tissue engineering. J Mater Sci Mater Med 16:1099–1104
55. Uebersax L et al (2006) Effect of scaffold design on bone morphology in vitro. Tissue Eng 12(12):3417–3429
56. Wang M (2003) Developing bioactive materials for tissue replacement. Biomaterials 24(13):2161–2175
57. Webster TJ et al (1999) Osteoblasts adhesion on nanophase ceramics. Biomaterials 20:1221–1227
58. Zhang R et al (1999) Poly(α-hydroxyl acids)/hydroxyapatite porous composites for bone-tis-sue engineering. Preparation and morphology. J Biomed Mater Res 44:446–455
59. Zhang R et al (1999) Porous poly(L-lactic acid)/apatite composites created by biomimetic process. J Biomed Mater Res 45(3):285–293

Chapter 2
Molecular Scissors: From Biomaterials Implant to Tissue Remodeling

Tudor Petreus, Iulian Antoniac, Paul Sirbu, and Carmen Elena Cotrutz

2.1 Introduction

As long as healthcare system improves and directs itself toward tissue substitution or replacement, all special materials with biomedical applications will become milestones for clinical and laboratory investigations [30, 56, 114]. We are now living in a dynamic world that is not yet prepared for a "*cyborg*" concept but still dreams about perfect tissue engineering, resulting in organ cloning. Thus, our age belongs to biomaterials and their interaction with living tissues. Tissue-implant interactions are well documented in the last decade, especially those regarding mobile cells in the blood, vascular endothelial cells and cells of various connectives [6]. All type of implantations, especially for non-resorptive materials (metallic, hard insoluble polymers) [99] involve tissue trauma, which induces an inflammatory response, followed by wound healing reaction (angiogenesis, fibroblast activation) and extracellular matrix (ECM) remodeling [65]. At the same time, any material intended for implantation should not impede cell development, growth or multiplication, in close vicinity or at certain distance. Moreover, all high performance implanted materials should limit the extension of inflammatory reactions and also should promote tissue remodeling toward a functional status [112, 113, 116].

T. Petreus (✉) • P. Sirbu • C.E. Cotrutz
Cell Biology Department, G.T. Popa University of Medicine and Pharmacy,
Universitatii Street 16, Iasi 700115, Romania
e-mail: biotudor@gmail.com; pdsirbu@yahoo.com; cotrutz@yahoo.com

I. Antoniac
University Politehnica of Bucharest, Spl. Independentei 313, Sect.6, Bucharest, Romania
e-mail: antoniac.iulian@gmail.com

I. Antoniac (ed.), *Biologically Responsive Biomaterials for Tissue Engineering*,
Springer Series in Biomaterials Science and Engineering 1,
DOI 10.1007/978-1-4614-4328-5_2, © Springer Science+Business Media New York 2013

Most of the implanted biomaterials, mainly bone prostheses lead to an overstated healing reaction as is fibrosis [112]. ECM production can also induce increased cell productivity. It is well documented, even if still poorly understood, that inflammatory processes may influence good biocompatibility, while tissue remodeling following inflammation plays an essential role in accurate organ function restoration [44, 112]. The ultimate goal for an implantable biomaterial should be the bioactivity, the capacity to support tissue regeneration and few or lack of fibrotic tissue.

An implanted biomaterial interacts not only with mobile cells or physiological body fluids but also with ECM. Almost all tissues (except epithelial tissues) possess an abundant ECM, with various compositions. Many implant failures may be due to an impaired cellular response (including inflammation) but all cell behavior can be influenced by the chemical composition and the physical properties of the ECM [57, 116]. The ECM represents a complex network of proteins and polysaccharides spread between the cells and which may include many types of molecules secreted by cells [15]. More than universal biological glue, ECM also forms highly specialized structures such as cartilage, tendons, basal laminas, and bone and teeth with some forms of calcium phosphate crystals [41, 81, 107]. The quantitative variations of different types of matrix macromolecules determine the diversity of the matrix shapes, consistent with the tissue functions. Thus, in the epithelial tissues the ECM is limited in volume, but in the connective tissues its volume is much larger than the cell's volume.

ECM plays an active and complex role, influencing cell development, migration, and proliferation while stabilizing the shapes and cell metabolic functions. During the differentiation process the cells require specific components. Cells morphogenesis depends closely on the ECM fibers. The ECM components can bind growth factors and hormones offering signal abundance for neighbor cells. While different cells may secrete various elements of ECM, we may distinguish two types [13, 49, 60, 75]:

1. Polysaccharide chains of glycosaminoglycans, with a covalent bound to a protein, forming proteoglycans.
2. Fibrillar proteins, which could be either structural (collagens and elastins) or adhesive (fibronectin and laminin).

Glycosaminoglycans and proteoglycans molecules from connective tissues form a highly hydrated gel-like essential substance in which collagen fibers are embedded. These collagen fibers strengthen up and help to organize the ECM while polysaccharide gel gives resistance to pressure. At the same time, the aqueous phase of the polysaccharide gel allows a fast nutrients, metabolites, and hormone diffusion between blood and tissues. Also, the elastin fibers enhance the ECM-specific elasticity, while some proteins as fibronectin are responsible for fibroblast or other cells' attachment to ECM in the connective tissues. For epithelial cells, laminin connects them to basement membrane (BM) [12, 47].

2.2 Extracellular Matrix Tailoring by Specific Means

Controlled interactions between ECM and mature differentiated cells may induce important changes in cells' behavior following specific signaling pathways [25, 58, 67]. Matrix tailoring supposes not only pure ECM dissection and cleavage for its main molecules but also subtle changes in ECM composition, which may alter the activity of the growth factors or of cells' surface receptors. In order for the essential biological processes to develop, ECM must be degraded in a controlled manner in order to allow cell displacement during embryonic development, tissue remodeling, and tissue repair [1, 85]. However, the same process of ECM degradation, if exceeded by the impairment of natural inhibitors, participates as innocent bystander or by active facilitator to growth, invasion, and metastasis of malignant cells. These processes are directly mediated by Matrix Metalloproteinases (MMPs), which we may consider to act as true molecular scissors for ECM.

2.3 Proteases and MMPs Family

All MMPs are included in the main class of proteases. These proteases include exo- and endopeptidases (also called proteinases). These proteinases can be divided in four classes according to the catalytic site and active center: serin/threonin proteinases, cysteine proteinases, aspartic proteinases, and metalloproteinases [20, 39, 71, 88]. Folding similarities are dividing metalloproteinases in clans while an evolutionary classification divides them in families. In metalloproteinases class we may include 8 clans and 40 families [121]. MMP members are included in the MB clan, characterized by a specific three His residues from which the third is bound to the Zn ion [24]. In this clan we can observe two major families (M12 and M10), each with two subfamilies, astacins and reprolysins for the first family and seralysins and matrixins (MMP) for the second. MMPs are gathered together in this subfamily of zinc and calcium-containing endoproteases that have been traditionally characterized by their collective ability to degrade all components of the ECM [121]. These enzymes are supposed to regulate the homeostasis in various tissues under the direct control of tissue inhibitor of metalloproteinases (TIMPs), which bind to and inhibit the activity of MMPs [19, 43]. Accordingly, an imbalance between MMPs and TIMPs can lead to a variety of pathological states, such as metastasis of cancer [29], or diseases including rheumatoid arthritis and multiple sclerosis [32]. At least 28 members of this enzyme family, which demonstrate significant sequence homology, have been reported [19, 22, 87]. Following functional criteria, they can be classified as collagenases (MMP-1, -8, -13, and -18), gelatinases (MMP-2 and -9), stromelysins (MMP-3, -10, and -11), matrilysins (MMP-7 and -26), enamelysin (MMP-20), metalloelastase (MMP-12), and membrane-type MMPs (MMP-14, -15, -16, -17, -24, and -25) [121]. Part of the newly discovered MMPs (after 1996) remain in a separate group, called "others" (MMP-19, -21, -23, -27, -28) while classification undergoes permanent changes, following new discoveries regarding their functionality [45, 83, 119]. For more than 30 years MMPs have been seen as promising targets for the treatment of the above-mentioned diseases because collagenase (MMP-1), gelatinases (MMP-2 and -9), and stromelysin-I (MMP-3) have been shown to play a key role in cancer invasion and metastasis [8, 19, 23, 82]. A large number of succinyl hydroxamates and sulfonamide hydroxamates have been reported as MMP inhibitors [55, 62, 106, 125]. Accordingly inhibitors of the gelatinases may have therapeutic potential for angiogenesis and/or tumor metastasis [28, 70, 90, 110].

Some authors [8, 16, 88] have nominated these enzymes as metzyncins due to the fact that they contain a Met residue in a carboxylic area near catalytic Zn ion. In this area, the polypeptidic chain of the catalytic site is bending and this bend influences the catalytic pocket activity. All and each detail regarding the catalytic site is important regarding the action of natural and synthetic inhibitors that are able to modulate MMP activity, regarding ECM degradation or tissue remodeling. In 1995, Krane [68] has formulated the principles for an MMP to be considered as an active factor for remodeling processes:

- Remodeling may be blocked by a drug or specific antibody directed against that specific MMP.
- Remodeling process may be reproduced by MMP over-expression on transgenic animals.
- Remodeling process may be interfered by specific MMP gene deletion.
- Spontaneous mutations can be identified and phenotypic characterization realized.
- Mutations can be induced in genes that reproduce the remodeling processes.

Thus, modulation of MMP activity appears to become essential in remodeling control, even in implant adaptation to host.

The far most important domain of the MMPs is the catalytic one [93, 120]. However we cannot ignore the other domains of this enzyme that include two Zn ions (one catalytic and one structural) and 2–3 Ca ions (Fig. 2.1). Starting from the N-terminus, we distinguish the following: (1) the signal peptide (12–20 hydrophonic amino acid residues). MMP-17 is lacking this domain; (2) pro-peptide includes around 80 residues and a highly conserved region toward the C-terminus; this end includes the Cys residue found in all MMPs, responsible for catalytic Zn contact and the stability for the zymogenic (inactive) form of the enzyme; (3) furin-cleavage domain includes around nine residues leading to furin cleavage; Only MMP 11, 14, 15, 16, and 17 include this domain while the other MMPs are cleaved by extracellular proteases, exposing the Zn ion and inducing autolytic cleavage for the remaining peptide; (4) catalytic domain, including approximately 160–170 residues, contains the Zn and Ca ions; (5) the 54-amino acid C-terminal region contains the catalytic Zn; (6) three repetitive fibronectin-like domains (mainly in MMP-2 and -9, serving for gelatin binding); (7) flexion region connects the catalytic domain to the next hemopexinic domain; (8) hemopexinic domain, including around 200 residues with four repeats that simulate hemopexin and vitronectin; this domain is involved in the recognition of the specific natural inhibitors (TIMPs); (9) membrane insertion domain, mainly for membrane type matrix metalloproteinases (MT-MMPs) with a length of 80–100 residues.

Fig. 2.1 Gelatinase A (MMP-2) catalytic domain, indicating the Zn ions coordinated by three histidines each and three Ca ions

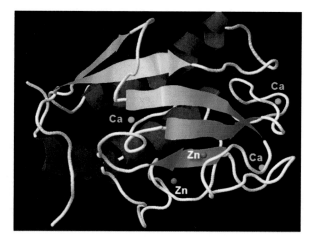

2.4   MMPs' Activity Depends on Catalytic Domain Structure and Behavior

Modulation of MMP activity depends directly on the specific inhibition or modulation of the catalytic domain. This inhibition depends on the intimate structure of the catalytic domain of the MMPs (which is common to almost all metalloproteinases) [93].

The catalytic domain of the MMPs exhibit the shape of an oblate ellipsoid. In the "standard" orientation, as depicted in most papers regarding MMPs [10, 11, 54] as preferred for this enzyme display, a small active-site cleft notched into the ellipsoid surface extends horizontally across the domain to bind peptide substrates from left to right. The MMP active site is composed of two separate regions: a groove at the protein surface surrounding the protruded catalytic Zn ion and a specificity site, nominated as S1′; the latter shows a higher degree of variability among the MMP in the family. Inside the groove, all tested synthetic inhibitors adopt extended conformations, express hydrogen bonds with the enzyme, and provide support for Zn ion coordination (see Fig. 2.1). Most current pharmacologic interventions have focused on development of MMP synthetic inhibitors that target MMPs in the extracellular space. MMP synthetic inhibitors are included into four categories: peptidomimetics, nonpeptidomimetics, tetracycline derivatives, and bisphophonates [55, 105]. Peptidomimetic inhibitors mimic the structure of collagen (a common substrate of MMPs) at the MMP cleavage site [51]. These compounds function as competitive inhibitors and chelate the catalytic zinc atom. Depending on the group that binds and chelates the zinc atom, peptidomimetic inhibitors for MMPs can be further subdivided into hydroxyamates, carboxylates, aminocarboxylates, sulfhydryls, and phosphoric acid derivatives. The hydroxamate MMPIs (batimastat, BB-1101 and marimastat) are broad spectrum inhibitors with significant activity in several animal models. Batimastat was the first MMPI to enter clinical trials but its utility was limited due to poor water solubility which required intraperitoneal administration

[69, 94]. Marimastat is a broad spectrum synthetic MMP inhibitor [63, 108] acting on MMP-1, -2, -3, -7 and -9—with good oral bioavailability that was studied in phase II and III clinical trials in many tumor types [40, 117].

The collection of universal chemical properties that characterizes the specific action of a ligand in the active site of a three-dimensional conformational model of a molecule is called a pharmacophore. The properties of the active catalytic site in an MMP are figured by the presence of hydrogen bridges, electrostatic interactions with an alleged inhibitor or hydrophobic areas. Pharmacophore modeling by computational processing becomes a universal, comprehensive and editable process. The inhibitor-catalytic site selectivity can be adjusted by adding or omitting some characteristics. Our group has used the LigandScout 2.0 software [123, 124] in order to align pharmacophores with important ligand molecules, based on their properties, in arbitrary combinations. The alignment of the two elements is realized by pairing only, regardless of the number of aligned elements. Thus, it is required to define a structure as marker element. In order to define the pharmacophores, from the 21 MMP-inhibitor complexes we have chosen nine files that are crystallographically defined in PDB databank: 1eub, 1fls, 1xuc, 1xud, 1xur, 1you, 1ztq, 456c, and 830c, respectively. We have chosen these complexes with experimentally defined affinity constants for a further usage of the pharmacophores, based on these determined complexes. The pharmacophores represent important filters regarding in silico evaluation of new MMP inhibitors. For automatic generation of MMP pharmacophore models we have imported the selected PDB files in LigandScout application. Due to the fact that PDB files do not include information regarding atom hybridization status and bond type, LigandScout application uses a complex algorithm to analyze the ligand structure in order to assign the bond type according to molecular geometry. Thus, it is recommended a manual check for the automatic deduction for the ligand structure. The key structure for the inhibitors CGS 27023 and WAY-151693 in the complex structure 1eub and 1fls respectively in PDB database is the isopropyl substituent and the basic 3-pyridyl substituent. Zn coordination is insured by the hydroxamic group while this inhibitor is part of the sulfonamide hydroxamates inhibitor family [48]. The non-hydroxamic inhibitor WAY-170523 in the complex 1ztq shows the replacement of the hydroxamic acid group with a carboxylate group; the latter is not Zn chelating but the inhibitory action is due to correct positioning of the benzofuran moiety of the P1' group by a biphenyl P1' linker that fills the hydrophobic S1' tunnel [94].

Another inhibitor, found in the 1xuc, 1xud, and 1xur complex is represented by pyrimidine dicarboxamide [17]. This inhibitor binds deeply in the S1' pocket and the pyridyl substitutions do not approach the catalytic zinc more than 5.5 Å. This inhibitor shows a bent conformation with the pyridyl substitutions close to Leu218.

For the pyrimidinetrione-based inhibitors of MMP13 in 1you PDB file, the aryloxyaryl ether fits in the S1 pocket while the pyrimidinetrione binds to the active zinc in the catalytic site [4, 14, 33, 97]. The three dimensional alignment of two inhibitors shown a score of around 77 from 100, that is remarkable for two different inhibitors in two different crystallographic determined structures.

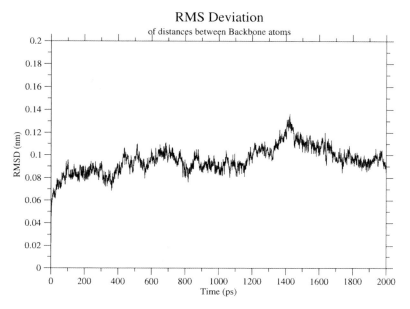

Fig. 2.2 Root mean square displacement (RMSD) for backbone atoms on the 1QIB file, representing the catalytic site of MMP-2, during a 2 ns molecular dynamic simulation

It is demonstrated that various MMPs exhibit different selectivity for many ECM proteins. Thus, it is important to understand such substrate selectivity to develop new synthetic MMP inhibitors. We have also shown that not only the Zn ion coordination in the P site is important, but also the hydrophobicity of S1 tunnel can be a step in further computer design for potent inhibitor or enzyme modulation factors.

The stability of the catalytic domain was subject for many studies [4, 26, 27] including our group investigation [92]. In order to appreciate this stability, we have performed molecular dynamics (MD) simulations of the catalytic domain of a known matrix metalloproteinase (MMP-2) for a 2 ns duration, in the absence of the substrate or a known inhibitor, starting from Protein Data Bank published data (ID-1QIB) [9]. We have chosen to the MMP2 (gelatinase A) catalytic domain, while it is the best characterized crystallographically. We have evaluated the conformational stability for MMP2 catalytic site during MD simulation.

In order to describe the variation of atom position into the simulated structures by comparison with the experimental defined conformation, it is necessary to plot the Root Mean Square Displacement (RMSD) over time (Fig. 2.2). We have tried to assess the stable states of the investigated by monitoring the RMSD of the protein backbone or only for the C-α carbon atoms along the trajectory, compared to the structure before starting the simulation. Literature data [18, 26, 100] show that RMSD values ranged between 0.1 nm and 0.3 nm, if constant, are representing a

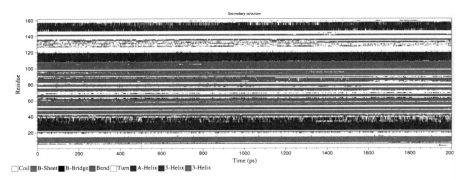

Fig. 2.3 Conservation of the secondary structure of the MMP2 catalytic site during the 2 ns MD simulation

conformational stability status for the simulated molecule. We have isolated the obtained trajectories for the catalytic domain of MMP-2 in the 1QIB file and we have analyzed the evolution of the secondary structure within simulation time. The domains of the catalytic site showed a good conservation over simulation time (Fig. 2.3).

By performing a Root Mean Square Fluctuation (RMSF) analysis, we have compared the NMR starting structure of the MMP2 catalytic site with the structures obtained by molecular dynamics simulations and we have noticed a high stability for this domain. The catalytic site stability was witnessed by the reduced fluctuations for the catalytic Zn area. However we have noticed some fluctuations, mainly in some loop regions near the catalytic site of the MMP, in 192–250 region of the MMP2 PDB file (394–449 on drawing) (Fig. 2.4).

Protein conformational changes are important mainly for enzymes and their activation status. These conformational changes are part of the kinetic states for proteins and include the variability of the secondary and tertiary structure and thus the global atomic movement in the investigated protein-enzyme. At the same time, side chain fluctuations and loop movements are characterizing the kinetic states of the observed substates in molecular dynamics simulations. MMP-2 shows a remarkable stability for the considered catalytic site and the presence of two substates. The analysis of free Gibbs energy landscape for MMP2 catalytic site showed a native state with two substates separated by less than 1KT (Fig. 2.5).

Thus, we have concluded that the dynamic signature for the MMP2 catalytic site is a native state with two substates, in close vicinity. We have also studied the secondary structure conservation, root mean square deviation (RMSD), and fluctuation cluster analysis for a full structured description of the crystal structure of the MMP-2 catalytic site. Based on these results, we can thus define new tricks for strategies regarding MMP inhibitors design. The role of the atomic position in the catalytic site for MMPs appear to be essential regarding to further simulations for conceiving a rather modulating inhibitor for these enzymes.

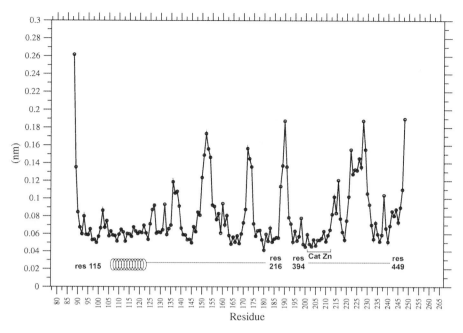

Fig. 2.4 Root mean square fluctuation (RMSF) for MMP2 catalytic site in 1QIB file

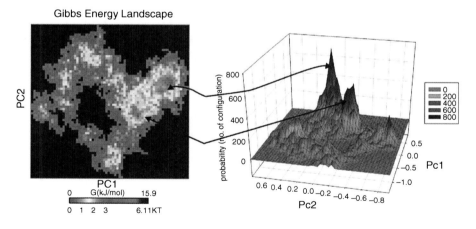

Fig. 2.5 Free Gibbs energy landscape for MMP2 catalytic site with two substates separated by less than 1KT

2.5   MMPs Natural Inhibitors

All MMPs are subject to natural inhibition by specific proteins called tissue inhibitors of matrix metalloproteinases (TIMPs). TIMPs bind to MMP in 1:1 or 2:2 ratios. TIMP-1 represents a more efficient inhibitor for MMP-1, MMP-3, and MMP-9 than

TIMP-2 [38, 79, 127]. Some authors [127] observed that TIMP-2 has a double effect on MMP-2 since MT-MMP-mediated proMMP-2 activation requires a tiny amount of TIMP-2 to make activation progress, whereas a higher concentration of TIMP-2 inhibits MMP-2 [64]. TIMP-3 may inhibit MMP-2 and MMP-9 [122], and TIMP-4 is a good inhibitor for all MMPs [109]. Tissue levels for TIMPs in ECM are largely exceeding MMP levels in the same areas. TIMPs act also as growth hormone and inductor for apoptosis (mainly TIMP3). TIMPs transcription process is regulated by cytokines and growth factors, involved in inflammation and remodeling processes. Natural inhibitors for MMP may represent important performers in this play. Some authors have described the close relation between MMP-2-TIMP-2 and MMP-9-TIMP-1 enzyme-inhibitor pairs [86]. Interactions between these pairs are indeed very complex and widely investigated from various points of view, by different experimental means [102] and for different purposes. TIMP-2 is involved in MMP-2 activation at the cell surface by forming a complex with MT1-MMP. Thus, TIMP-2 can play a dual role during interaction with MMP-2. The biochemical relationship between TIMP-2 and MMP-2 may induce surface-mediated enzyme activation, while simple interaction TIMP-2–MMP-2 can determine enzyme inhibition. Our group has also investigated the colocalization and complex relationship between MMP-2, MMP-9 and TIMP-2, by molecular docking studies [92] to evaluate the docking options for these enzymes and inhibitor. While we have found described in the literature only the interaction MMP-9–TIMP-1, we have investigated the possible interaction between MMP-9 and TIMP-2 by molecular docking means. Thus we have evaluated first the optimal positions of the TIMP-2 around MMP-2 molecule and then, using a similitude procedure, we did the same thing for MMP-9–TIMP-2 complexes, for which there were no data regarding direct interaction. The investigated docking molecular models were MMP2–TIMP2 and MMP-9–TIMP-2. For the MMP-2–TIMP-2 enzyme-natural inhibitor complex, we have selected the 1gxd file from Protein Data Bank (PDB) [9] from which there were picked individually the molecules involved in this interaction, MMP-2 and TIMP-2. This selection was followed by an identification of the specific regions involved in the enzyme-inhibitor interactions, mainly the catalytic site and the hemopexinic site for MMP-2. This enzyme has the unique property of being able to bind successively two TIMP-2 molecules, either on hemopexinic site (playing a role in enzyme activation) or on the catalytic site (playing a role in enzyme inhibition). The docking process starts from two known extracted structures that form a complex (MMP2–TIMP2 in 1gxd file); the extraction of these proteins from the complex should not alter the initial structure of the complex; this procedure was then validated by superposition. We have checked this algorithm for this modeling software, while there is no available data regarding the interaction MMP-9–TIMP-2. Following rigid docking and filtering, we have a picture of the most probable positions for the inhibitor around the enzyme. The filtering procedure was performed for localizing the catalytic and the hemopexinic sites for TIMP-2 binding. The two resulted structures for MMP-2 and TIMP-2 (that occupies the best position near the catalytic site) were superposed over the crystallographic model represented in the 1gxd file from PDB database. Following superposition we have calculated also the Root Mean Square deviation

Fig. 2.6 Final refined complex final MMP2 (hemopexinic)—TIMP2, with the choice of the position with the most favorable score for docking

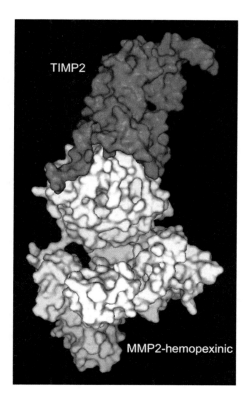

(RMSd), the distance between the initial atom coordinates in the crystallographic structure and the final coordinates of the partially altered structures by the modeling process. RMSd values of 2.30 Å indicated a correct superimposition and validated the method for further enzyme inhibitor pairs. After model refinement, we have represented the results with VMD viewer [53]. We have represented the final complexes MMP-2 (catalytic-refined)–TIMP-2 (Fig. 2.6) and MMP2 (hemopexinic-refined)–TIMP2 (Fig. 2.7).

MMP-9–TIMP-2 model raised some problems while PDB databases do not include yet a crystallographic model but only a theoretical one. The chosen model was 1LKG. In PDB database we have selected the hemopexinic domains (1ITV-dimeric form) and catalytic (1L6J) [9]. The 1L6J file contains the whole MMP9 sequence from which we have selected the catalytic domain for the experimental model. The domains were superimposed over the theoretical model for MMP-9 (Fig. 2.8).

The docking procedure for the MMP9–TIMP2 complex determined 10,000 possible variants that have been filtered and selected from the minimal coupling energy then graphically represented. Following this molecular docking study we have concluded that the natural inhibitor TIMP2 has a higher affinity for the catalytic

Fig. 2.7 Final refined
complex MMP2 (catalytic)—
TIMP2, with the choice of
the position with the most
favorable score for docking

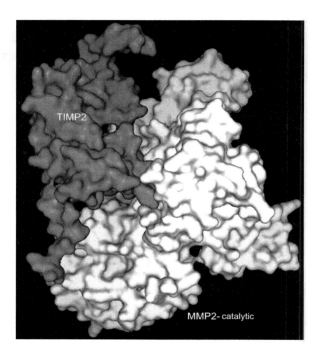

Fig. 2.8 Alignment for
hemopexinic and catalytic
domains with the theoretical
model of MMP9

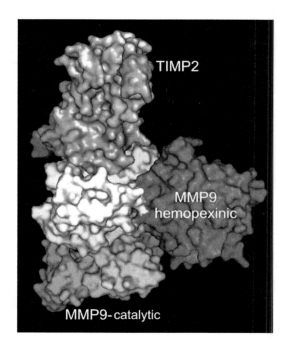

domain than for the hemopexinic domain of MMP. Regarding the activation mechanism, these data correspond to the well-known experimental results [7, 31, 77], that sustain that initially; TIMP2 forms an activation complex with the hemopexinic domain in MMP2.

Even if has a higher affinity for TIMP2, the catalytic domain cannot bind it while the specific binding site is inactive in the pro-form of the enzyme. Following MMP2 activation and removal of the pro-domain of the enzyme, the catalytic site becomes available and the favorable free binding energy will induce either the substrate binding or the natural inhibitor TIMP2. We have confirmed that MMP2 can consecutively bind two TIMP2 molecules, one for the activation complex at the hemopexinic domain level and the other at the catalytic site, inducing inhibition. For the MMP9–TIMP2 complex, the affinity for the hemopexinic and the catalytic site was almost the same. Thus, the activation-inhibition model in this triad could be [91]: MMP2 together with TIMP2 and MT-MMP1 forms an activation complex that activates MMP2; the activated MMP2 attacks the specific substrate. At the same time TIMP2 will inhibit MMP9 at the catalytic site while the hemopexinic site is already coupled with TIMP1 that participates to its activation.

2.6 MMPs in Wound Repair

MMPs participate to all wound repair steps following biomaterial implantation (Fig. 2.9) [34, 73].

Inflammation step is mediated by cytokines and chemokines incoming from resident cells in the tissue to which implantation is addressed (epithelial or endothelial cells, fibroblasts). MMPs can activate the inflammatory mediators at cell surface either by direct processing (cleavage of a pro-domain) or even by degradation, thus inhibiting pro-inflammatory signals. MMPs are also able to cleave molecular domains of the intercellular or cell-matrix junctions in epithelia in order to promote re-epithelization. Moreover, MMPs are involved in scar tissue remodeling either by direct proteolytic cleavage of collagen molecules or indirect, by influencing cell behavior. ECM processing is complete at the end of wound repair process but is also involved in regulation of the inflammatory process. MMPs are playing key regulation factor roles regarding multiple processes in tissue repair [52, 98].

MMPs are also playing important roles in inflammatory processes. As we have already emphasized, inflammation is one of the first processes at the biomaterial implant site. Inflammation processes are associated to leukocyte influx and activation and are influenced by antimicrobial peptides, lipid mediators, receptors, chemokines, and cytokines, fragments derived from ECM. Production and activation of these factors is controlled by effectors that include MMPs. Specific MMPs derived from epithelial cells are regulating numerous steps during inflammation, as leukocyte trans-epithelial migration, activity, and distribution of chemokines [21]. Inflammatory cells are expressing MMPs as all other epithelial

band desmosome

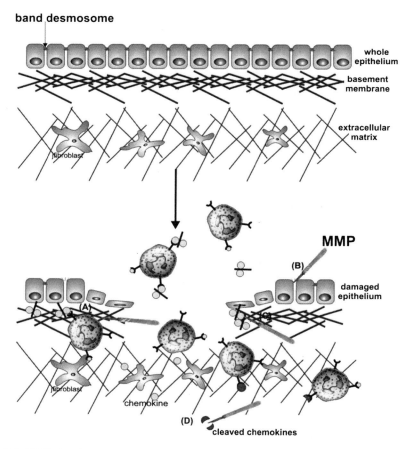

Fig. 2.9 MMPs involvement in healing steps. (**a**) Cell migration is facilitated by MMPs by degradation of the adhesive and ECM proteins; (**b**) Re-epithelization—adhesion proteins are cleaved by MMPs and cells involved in healing are promoted; (**c**) Leukocyte influx—chemokines and proteoglycans are cleaved by MMPs; (**d**) Inflammation step; MMPs are processing multiple chemokines

and stromal cells in wounds. In wound repair processes we may observe the involvement of MMP-1, 2, 3, 7, 9, 10, or 28. Many of these MMPs may regulate chemokine activity by direct proteolysis or by generating chemokine gradients. Chemokines are divided in subfamilies based on N-terminal cysteinic residues and play specific roles in attracting the leukocyte flow toward inflammation site. CC chemokines are characterized by two neighbor cysteinic residues toward N-terminus and are responsible for monocyte chemotaxis. CXC chemokines show another amino acid between cysteine residues and are responsible for neutrophil attraction toward inflammation site. Chemokine effects are mediated by binding to 7-domain-transmembrane G protein-coupled receptors (7-TM-GPCRs). MMP-controlled chemokine cleavage induces a reduction of chemokine activity.

This mechanism characterizes CC-chemokines that are producing antagonists for the receptors that will inhibit signaling pathways, following MMP processing [126]. It appears that chemokines CCL-2, 5, 7, 8, and 13 are more efficiently and frequently activated by MMP-1 and -3. CXC chemokines show variable response following cleavage by MMPs. Some chemokines are resistant to MMP processing (CXCL1, -2 and -3) while others are processed by multiple MMPs (CXCL5, -12 and interleukin-8). Human CXCL8 [35] is processed by MMP-8 and -9 and induce their activation. CXCL5 (LIX) chemokine [66] is processed by MMP 1, 2, 8, 9, and 13. This chemokine processing determines an increase of its activity that promotes an accelerated recruitment of the inflammatory cells. CXCL12 (SDF1) chemokine is also processed by multiple MMPs (1, 2, 3, 9, 13, 14) and its processing induces a decrease of the chemokine activity [42]. From all MMPs, it appears that MMP-1, -3, and -9 are the most potent agents for modulating chemokine signaling activity.

Other MMPs (disintegrins) are able to cleave CXCL16 at cell surface, allowing binding to specific receptor and regulating the T-cell activity in wound repair [2, 36, 111]. MMPs are also required to define chemotactic gradient. MMP-7 represents a key factor for regulating transepithelial neutrophils migration. Injured epithelial cells are expressing CXCL1 that bind to heparan-sulfate glycosaminoglycan chains and to syndecan-1, thus generating a chemical gradient required for activated neutrophils. In the absence of MMP-7, neutrophils are not migrating anymore and are sequestrated into intercellular space. MMPs and disintegrins ADAM's (*A Disintegrin And Metalloproteinase*) play an important role in diapedesis [61]. ADAM-28 is expressed by lymphocytes in the peripheral blood and is located as membrane coupled secreted enzymes [50]. It may bind the selectin glycoprotein at lymphocyte surface thus activating P-selectin binding on endothelial cells. MMPs are able to cleave occludins from tight junctions and VE-cadherins (vascular endothelial) from adhesion junctions, being involved in endothelial permeability [96]. Inflammation represents a main step in wound healing and in tissue early response to implant. Peri-implant reactivity following biomaterial implantation into soft tissues involves the follow-up for inflammatory reactions, necrosis, granuloma presence dystrophy or calcification reactions around implanted tissue [5]. These reactions are induced and depend on the germ contamination of the implanted materials. Inflammatory reaction is notable from 2 to 4 weeks, being usually moderate and limited to a very thin area around the implant, regardless the biomaterial physical properties. Between 4 and 12 weeks the fibrous capsule shows two separate areas: a layer of inflammatory cells with a moderate inflammatory response more reduced than the previous interval and a fibrous area formed by a collagen network and fibroblasts. After 26 weeks, only a thin fibrous capsule remains, where macrophages and inflammatory cells are missing. We can describe new formed capillaries between the fibrous capsule and the surrounding tissues [72].

Our group has performed various biomaterials testing in vivo by subcutaneous rat implantation followed by analysis of peri-implant tissues in order to detect inflammatory processes and their consequences. Following some collagenated

Fig. 2.10 MMP8 positive
lymphocytes, detected by
immunohistochemistry,
around small capillaries

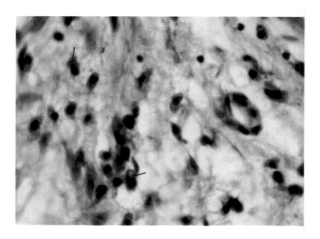

networked biomaterial implant, at 16 weeks following implantation; we have
observed a specific periimplant fine fibrous capsule [84]. In this capsule we have
detected, by immunohistochemical means, the presence of MMP-8 in few lymphocytes, close to fine capillaries (Fig. 2.10).

A specific role is played by MMPs during proliferative step of the wound repair
process [34]. Beside proliferation and epithelization induced by macrophages that
are stimulating the fibroblasts using interleukin-1 (IL-1) and keratinocyte growth
factor -2 (KGF-2), the ECM matrix synthesis is augmented by the same macrophages. They are initiating the formation of platelet-derived growth factor (PDGF)
and tumor necrosis factor α (TNF-α). Macrophages are also delivering proteases
that are activating the transforming growth factor β (TGF-β). The later stimulates
collagen production by fibroblasts. It also prevents collagen degradation by stimulating TIMP secretion, induces fibronectin secretion and increases the expression of
integrin receptors to allow cell adhesion to ECM [37, 89].

2.7 MMPs in Epithelial Repair

A main process in wound repair is represented by re-epithelization in which epithelial cells are regenerated on a denudated surface. Re-epithelization supposes the loss
of intercellular and cell-matrix contacts near the wound edges while cells are starting to migrate toward the lesion. May MMPs are involved in this wound repair step,
mainly MMP-1, -3, -7, 9, -10, -14 and -28 [3, 74, 89].

MMP-1 (collagenase-1) is found in cutaneous wounds in re-epithelization step
and its expression ceases while wound repair is complete. In the wounded tegument,
keratinocyte migration from the basement membrane meets a rich-collagen dermal
matrix, and binding to this element of the ECM (which is intermediated by α2β1
integrin) stimulates MMP-1 expression and also for other MMP genes. MMP-1

facilitates keratinocyte migration in dermal matrix by weakening the interactions between collagen and integrins [46]. MMP-10 (stromelysin-2) is colocalized with MMP-1 while MMP-3 (stromelysin-1) is localized in the cells behind the migration wave. Lack of superposition between MMP-3 and -10 (even if they are MMPs with similar structure) suggests that MMPs are playing various roles in re-epithelization processes [103]. MMP-7 regulates re-epithelization by cleavage of E-cadherins in adherence junctions, thus facilitating cell migration. MMP-28 (epilysin) is expressed by keratinocytes localized distally from injured area and lacks from migrating keratinocytes. This suggests a possible role in reorganization of the basement membrane or in cleavage of the adhesion molecules, releasing new cells for the migration wave [103]. MMP-9 (gelatinase B) is also involved in re-epithelization process following local injury [76, 78, 95]. Epidermal growth factor (EGF) and hepatocyte growth factor (HGF) stimulate the keratinocyte migrating ability. Cell migration depends on the MMP-9 activation and it is modulated by TIMPs or by synthetic specific inhibitors. MMP-10 is expressed by epithelial cells in the migrating cell wave and its over-expression may induce aberrant cell migration at wound edge level, reducing the abundance of laminin-5 mainly by direct cleavage. Laminin-5 decreased levels are associated with impairment of the cell-matrix signaling pathways and increase the number of keratinocytes that undergo apoptosis. Proliferation of the epithelial cells behind cell migration wave in the injured area is regulated by MMP-14. It appears that this MMP activity is directly related to keratinocyte growth factor receptors processing during epithelial injury repair. Thus, wound healing is similar to tissue morphogenesis processes that include cell migration, proliferation, differentiation, and cell death [59, 104].

2.8 MMP Involvement in Resolution Step of Wound Healing

It appears that MMP-3 may influence reorganization of actin filaments in dermal cultured fibroblasts that initiates early injured area contraction. It appears that ECM remodeling processes depend on various MMPs as MMP-1, -3, -13, and -14, enzymes that are able to cleave collagen. Also, MMP-7 may process syndecan or elastin. Collagen remodeling that includes degradation of preexistent fibrils and synthesis of new ones represent the main process in wound resolution. The enzymes involved in this process appear to be MMP-2 and -9 together with MMP-3 that are localized at epithelial–stromal interface, behind the epithelial cell migration wave, that suggests involvement in stromal and basement membrane remodeling [85].

2.9 TIMPs in Wound Repair

Inflammatory response is mediated by multiple cytokines and chemokines. Among cytokines involved in acute inflammation, TNF-alpha is activated by cleavage at membrane level by the specific conversion enzyme. One effect on inflammatory

cells (as monocytes) is to stimulate MMP-9 expression by NFkB activation and
MAP-kinase signaling pathway. MMP-9 over-expression may impair normal
healing and cicatrization process [80, 104, 115, 118].

TIMP-3 is one of the primary inhibitors for ADAM-17 and is involved in
inflammation regulation. While TIMP-3 is missing or secreted at low concentrations, ADAM-17 activity is accentuated, inducing TNF alpha secretion and an
increase of the inflammatory fluid volume. Interaction between TIMP-3 and
ADAM-17 plays an important role in recruiting and redirecting leukocytes.
Vascular cell adhesion molecules (VCAM) are also involved in white blood
cells recruitment, and during inflammatory response they are cleaved at endothelial cell surface by ADAM-17. TIMP-3 excess blocks soluble VCAM production while the decreases of TIMP-3 levels are activating surface cleavage of
VCAM-1 [118].

TIMP-1 may be involved in leukocyte flowing and vascular permeability by regulating apoptotic processes in endothelial cells. TIMP-1 plays an antiapoptotic
effect in endothelial cells. It appears that natural inhibitors for MMPs, mainly
TIMP-3, are required to adjust inflammatory cell response, by regulating cytokine
signaling and cell adhesion receptors.

TIMP also regulates some steps in cell migration. Epithelial cell migration
depends on TIMP-1 levels and is impaired by its low concentrations. There are
significant differences between epithelial and stromal cell migration during
re-epithelization and granulation tissue formation, while TIMP balance these processes [34].

2.10 Evaluation of Immunohistochemical Detection of MMPs in Periimplant Tissues

Our group has performed various tests in order to detect the presence of some
MMPs (as gelatinases) in periimplant tissues. We have used simple immunohistochemical detection of MMP-2 and -9, together with the detection of the TIMP-1
and -2, by using a DAB detection system bound to specific primary antibodies
against MMPs and TIMPs. For a semiquantitative evaluation of the immunohistochemistry results, we have used the color deconvolution technique, imagined and
described by Ruifrok și Johnston [101]. The analysis of our results showed that
MMP-9 is present in various amounts (expressed in area percentage on the examined slides) of 33,010, 27,878, 15,857, 30,213, and 31,852 %, compared to the
normal witness slide (without implant) where the MMP-9 marked area was of
9,614 % only. Thus, MMP appear to be involved in ECM remodeling processes in
the peri-implant area. TIMP-1 expression is also increased in the peri-implant area,
showing that remodeling processes require the presence of the enzyme and its natural inhibitor at the same time.

2.11 Summary

Tissue-implant interactions are well documented in the last decade, especially those regarding mobile cells in the blood, vascular endothelial cells, and cells of various connective tissues. All types of implantation, especially for non-resorptive materials (metallic, hard insoluble polymers) involve tissue trauma, which induces an inflammatory response, followed by wound healing reaction (angiogenesis, fibroblast activation) and ECM remodeling. All high performance implanted materials should limit the extension of inflammatory reactions and also should promote tissue remodeling toward a functional status. Matrix tailoring supposes not only pure ECM dissection and cleavage for its main molecules but also subtle changes in ECM composition, which may alter the activity of the growth factors or of cells' surface receptors. In order for the essential biological processes to develop, ECM must be degraded in a controlled manner in order to allow cell displacement during embryonic development, tissue remodeling, and tissue repair. These processes are directly mediated by Matrix Metalloproteinases (MMPs), that we may consider to act as true molecular scissors for ECM. Modulation of MMP activity appears to become essential in remodeling control, even in implant adaptation to host. Modulation of MMP activity depends directly on the specific inhibition or modulation of the catalytic domain. This inhibition depends on the intimate structure of the catalytic domain of the MMPs (which is common to almost all metalloproteinases). Our group has performed various tests in order to detect the presence of some MMPs (as gelatinases) in periimplant tissues, including molecular modeling studies together with immunohistochemical investigations.

References

1. Abreu T, Silva G (2009) Cell movement: new research trends. Nova Biomedical Books, New York
2. Agren MS, Jorgensen LN, Andersen M, Viljanto J, Gottrup F (1998) Matrix metalloproteinase 9 level predicts optimal collagen deposition during early wound repair in humans. Br J Surg 85(1):68–71
3. Arumugam S, Jang YC, Chen-Jensen C, Gibran NS, Isik FF (1999) Temporal activity of plasminogen activators and matrix metalloproteinases during cutaneous wound repair. Surgery 125(6):587–593
4. Aschi M, Bozzi A, Di Bartolomeo R, Petruzzelli R (2010) The role of disulfide bonds and N-terminus in the structural properties of hepcidins: insights from molecular dynamics simulations. Biopolymers 93(10):917–926
5. Atassi F (2002) Periimplant probing: positives and negatives. Implant Dent 11(4):356–362
6. Baier RE, Meenaghan MA, Hartman LC, Wirth JE, Flynn HE, Meyer AE et al (1988) Implant surface characteristics and tissue interaction. J Oral Implantol 13(4):594–606
7. Banyai L, Tordai H, Patthy L (1994) The gelatin-binding site of human 72 kDa type IV collagenase (gelatinase A). Biochem J 298(Pt 2):403–407
8. Baramova E, Foidart JM (1995) Matrix metalloproteinase family. Cell Biol Int 19(3):239–242

9. Bernstein FC, Koetzle TF, Williams GJ, Meyer EF Jr, Brice MD, Rodgers JR et al (1978) The Protein Data Bank: a computer-based archival file for macromolecular structures. Arch Biochem Biophys 185(2):584–591

10. Bertini I, Calderone V, Fragai M, Luchinat C, Mangani S, Terni B (2004) Crystal structure of the catalytic domain of human matrix metalloproteinase 10. J Mol Biol 336(3):707–716

11. Betz M, Huxley P, Davies SJ, Mushtaq Y, Pieper M, Tschesche H et al (1997) 18-A crystal structure of the catalytic domain of human neutrophil collagenase (matrix metalloproteinase-8) complexed with a peptidomimetic hydroxamate primed-side inhibitor with a distinct selectivity profile. Eur J Biochem 247(1):356–363

12. Bittar EE, Bittar N (1995) Cellular organelles and the extracellular matrix. JAI Press, Greenwich, CT

13. Black LD, Allen PG, Morris SM, Stone PJ, Suki B (2008) Mechanical and failure properties of extracellular matrix sheets as a function of structural protein composition. Biophys J 94(5):1916–1929

14. Blagg JA, Noe MC, Wolf-Gouveia LA, Reiter LA, Laird ER, Chang SP et al (2005) Potent pyrimidinetrione-based inhibitors of MMP-13 with enhanced selectivity over MMP-14. Bioorg Med Chem Lett 15(7):1807–1810

15. Bosman FT, Stamenkovic I (2003) Functional structure and composition of the extracellular matrix. J Pathol 200(4):423–428

16. Butler GS, Tam EM, Overall CM (2004) The canonical methionine 392 of matrix metalloproteinase 2 (gelatinase A) is not required for catalytic efficiency or structural integrity: probing the role of the methionine-turn in the metzincin metalloprotease superfamily. J Biol Chem 279(15):15615–15620

17. Carrascal N, Rizzo RC (2009) Calculation of binding free energies for non-zinc chelating pyrimidine dicarboxamide inhibitors with MMP-13. Bioorg Med Chem Lett 19(1):47–50

18. Chakrabarti B, Bairagya HR, Mallik P, Mukhopadhyay BP, Bera AK (2011) An insight to conserved water molecular dynamics of catalytic and structural $Zn(+2)$ ions in matrix metalloproteinase 13 of human. J Biomol Struct Dyn 28(4):503–516

19. Clark IM, Young DA, Rowan AD (2010) Matrix metalloproteinase protocols, 2nd edn. Humana, New York, NY

20. Conant K, Gottschall PE (2005) Matrix metalloproteinases in the central nervous system. Imperial College Press, London, Hackensack, NJ, Distributed by World Scientific

21. Cornwell KG, Landsman A, James KS (2009) Extracellular matrix biomaterials for soft tissue repair. Clin Podiatr Med Surg 26(4):507–523

22. Coussens LM, Tinkle CL, Hanahan D, Werb Z (2000) MMP-9 supplied by bone marrow-derived cells contributes to skin carcinogenesis. Cell 103(3):481–490

23. Curran S, Dundas SR, Buxton J, Leeman MF, Ramsay R, Murray GI (2004) Matrix metalloproteinase/tissue inhibitors of matrix metalloproteinase phenotype identifies poor prognosis colorectal cancers. Clin Cancer Res 10(24):8229–8234

24. Curran S, Murray GI (1999) Matrix metalloproteinases in tumour invasion and metastasis. J Pathol 189(3):300–308

25. Damsky CH, Werb Z (1992) Signal transduction by integrin receptors for extracellular matrix: cooperative processing of extracellular information. Curr Opin Cell Biol 4(5):772–781

26. Diaz N, Suarez D (2007) Molecular dynamics simulations of matrix metalloproteinase 2: role of the structural metal ions. Biochemistry 46(31):8943–8952

27. Diaz N, Suarez D (2008) Molecular dynamics simulations of the active matrix metalloproteinase-2: positioning of the N-terminal fragment and binding of a small peptide substrate. Proteins 72(1):50–61

28. Eck SM, Hoopes PJ, Petrella BL, Coon CI, Brinckerhoff CE (2009) Matrix metalloproteinase-1 promotes breast cancer angiogenesis and osteolysis in a novel in vivo model. Breast Cancer Res Treat 116(1):79–90

29. Egeblad M, Werb Z (2002) New functions for the matrix metalloproteinases in cancer progression. Nat Rev Cancer 2(3):161–174

30. El Haj AJ, Cartmell SH (2010) Bioreactors for bone tissue engineering. Proc Inst Mech Eng H 224(12):1523–1532
31. Ennis BW, Matrisian LM (1994) Matrix degrading metalloproteinases. J Neurooncol 18(2): 105–109
32. Fainardi E, Castellazzi M, Tamborino C, Trentini A, Manfrinato MC, Baldi E et al (2009) Potential relevance of cerebrospinal fluid and serum levels and intrathecal synthesis of active matrix metalloproteinase-2 (MMP-2) as markers of disease remission in patients with multiple sclerosis. Mult Scler 15(5):547–554
33. Freeman-Cook KD, Reiter LA, Noe MC, Antipas AS, Danley DE, Datta K et al (2007) Potent selective spiropyrrolidine pyrimidinetrione inhibitors of MMP-13. Bioorg Med Chem Lett 17(23):6529–6534
34. Gailit J, Clark RA (1994) Wound repair in the context of extracellular matrix. Curr Opin Cell Biol 6(5):717–725
35. Galvez BG, Genis L, Matias-Roman S, Oblander SA, Tryggvason K, Apte SS et al (2005) Membrane type 1-matrix metalloproteinase is regulated by chemokines monocyte-chemoattractant protein-1/ccl2 and interleukin-8/CXCL8 in endothelial cells during angiogenesis. J Biol Chem 280(2):1292–1298
36. Gapski R, Barr JL, Sarment DP, Layher MG, Socransky SS, Giannobile WV (2004) Effect of systemic matrix metalloproteinase inhibition on periodontal wound repair: a proof of concept trial. J Periodontol 75(3):441–452
37. Gapski R, Hasturk H, Van Dyke TE, Oringer RJ, Wang S, Braun TM et al (2009) Systemic MMP inhibition for periodontal wound repair: results of a multi-centre randomized-controlled clinical trial. J Clin Periodontol 36(2):149–156
38. Geisler S, Lichtinghagen R, Boker KH, Veh RW (1997) Differential distribution of five members of the matrix metalloproteinase family and one inhibitor (TIMP-1) in human liver and skin. Cell Tissue Res 289(1):173–183
39. Giaccone G, Soria J-C (2007) Targeted therapies in oncology. Informa Healthcare, New York
40. Goffin JR, Anderson IC, Supko JG, Eder JP Jr, Shapiro GI, Lynch TJ et al (2005) Phase I trial of the matrix metalloproteinase inhibitor marimastat combined with carboplatin and paclitaxel in patients with advanced non-small cell lung cancer. Clin Cancer Res 11(9):3417–3424
41. Gramoun A, Goto T, Nordstrom T, Rotstein OD, Grinstein S, Heersche JN et al (2010) Bone matrix proteins and extracellular acidification: potential co-regulators of osteoclast morphology. J Cell Biochem 111(2):350–361
42. Grassi F, Cristino S, Toneguzzi S, Piacentini A, Facchini A, Lisignoli G (2004) CXCL12 chemokine up-regulates bone resorption and MMP-9 release by human osteoclasts: CXCL12 levels are increased in synovial and bone tissue of rheumatoid arthritis patients. J Cell Physiol 199(2):244–251
43. Greenwald RA, Zucker S, Golub LM (1999) Inhibition of matrix metalloproteinases: therapeutic applications. New York Academy of Sciences, New York, NY
44. Gristina AG (1994) Implant failure and the immuno-incompetent fibro-inflammatory zone. Clin Orthop Relat Res (298):106–118
45. Gruber HE, Ingram JA, Hoelscher GL, Zinchenko N, Norton HJ, Hanley EN Jr (2009) Matrix metalloproteinase 28 a novel matrix metalloproteinase is constitutively expressed in human intervertebral disc tissue and is present in matrix of more degenerated discs. Arthritis Res Ther 11(6):R184
46. Gu Q, Wang D, Gao Y, Zhou J, Peng R, Cui Y et al (2002) Expression of MMP1 in surgical and radiation-impaired wound healing and its effects on the healing process. J Environ Pathol Toxicol Oncol 21(1):71–78
47. Hay ED (1991) Cell biology of extracellular matrix, 2nd edn. Plenum, New York
48. He H, Puerta DT, Cohen SM, Rodgers KR (2005) Structural and spectroscopic study of reactions between chelating zinc-binding groups and mimics of the matrix metalloproteinase and disintegrin metalloprotease catalytic sites: the coordination chemistry of metalloprotease inhibition. Inorg Chem 44(21):7431–7442

49. Hinds S, Bian W, Dennis RG, Bursac N (2011) The role of extracellular matrix composition in structure and function of bioengineered skeletal muscle. Biomaterials 32(14):3575–3583
50. Howard L, Zheng Y, Horrocks M, Maciewicz RA, Blobel C (2001) Catalytic activity of ADAM28. FEBS Lett 498(1):82–86
51. Van den Hu J, Steen PE, Houde M, Ilenchuk TT, Opdenakker G (2004) Inhibitors of gelatinase B/matrix metalloproteinase-9 activity comparison of a peptidomimetic and polyhistidine with single-chain derivatives of a neutralizing monoclonal antibody. Biochem Pharmacol 67(5):1001–1009
52. Hubbell J (2006) Matrix-bound growth factors in tissue repair. Swiss Med Wkly 136(25–26): 387–391
53. Humphrey W, Dalke A, Schulten K (1996) VMD: visual molecular dynamics. J Mol Graph 14(1):33–38, 27–38
54. Hyer S, Wei S, Brew K, Acharya KR (2007) Crystal structure of the catalytic domain of matrix metalloproteinase-1 in complex with the inhibitory domain of tissue inhibitor of metalloproteinase-1. J Biol Chem 282(1):364–371
55. Janusz MJ, Hookfin EB, Brown KK, Hsieh LC, Heitmeyer SA, Taiwo YO et al (2006) Comparison of the pharmacology of hydroxamate- and carboxylate-based matrix metalloproteinase inhibitors (MMPIs) for the treatment of osteoarthritis. Inflamm Res 55(2):60–65
56. Jayakumar P, Di Silvio L (2010) Osteoblasts in bone tissue engineering. Proc Inst Mech Eng H 224(12):1415–1440
57. Jones KS (2008) Effects of biomaterial-induced inflammation on fibrosis and rejection. Semin Immunol 20(2):130–136
58. Juliano RL, Haskill S (1993) Signal transduction from the extracellular matrix. J Cell Biol 120(3):577–585
59. Karim RB, Brito BL, Dutrieux RP, Lassance FP, Hage JJ (2006) MMP-2 assessment as an indicator of wound healing: a feasibility study. Adv Skin Wound Care 19(6):324–327
60. Kass L, Erler JT, Dembo M, Weaver VM (2007) Mammary epithelial cell: influence of extracellular matrix composition and organization during development and tumorigenesis. Int J Biochem Cell Biol 39(11):1987–1994
61. Kataoka H (2009) EGFR ligands and their signaling scissors ADAMs as new molecular targets for anticancer treatments. J Dermatol Sci 56(3):148–153
62. Khan OF, Jean-Francois J, Sefton MV (2010) MMP levels in the response to degradable implants in the presence of a hydroxamate-based matrix metalloproteinase sequestering biomaterial in vivo. J Biomed Mater Res A 93(4):1368–1379
63. Kimata M, Otani Y, Kubota T, Igarashi N, Yokoyama T, Wada N et al (2002) Matrix metalloproteinase inhibitor marimastat decreases peritoneal spread of gastric carcinoma in nude mice. Jpn J Cancer Res 93(7):834–841
64. Kinoshita T, Sato H, Okada A, Ohuchi E, Imai K, Okada Y et al (1998) TIMP-2 promotes activation of progelatinase A by membrane-type 1 matrix metalloproteinase immobilized on agarose beads. J Biol Chem 273(26):16098–16103
65. Kirkpatrick CJ, Krump-Konvalinkova V, Unger RE, Bittinger F, Otto M, Peters K (2002) Tissue response and biomaterial integration: the efficacy of in vitro methods. Biomol Eng 19(2–6):211–217
66. Koltsova EK, Ley K (2010) The mysterious ways of the chemokine CXCL5. Immunity 33(1):7–9
67. Kornberg L, Juliano RL (1992) Signal transduction from the extracellular matrix: the integrin-tyrosine kinase connection. Trends Pharmacol Sci 13(3):93–95
68. Krane SM (1995) Is collagenase (matrix metalloproteinase-1) necessary for bone and other connective tissue remodeling? Clin Orthop Relat Res (313):47–53
69. Kruger A, Soeltl R, Sopov I, Kopitz C, Arlt M, Magdolen V et al (2001) Hydroxamate-type matrix metalloproteinase inhibitor batimastat promotes liver metastasis. Cancer Res 61(4):1272–1275
70. Kurizaki T, Toi M, Tominaga T (1998) Relationship between matrix metalloproteinase expression and tumor angiogenesis in human breast carcinoma. Oncol Rep 5(3):673–677
71. Lagente V, Boichot E (2008) Matrix metalloproteinases in tissue remodelling and inflammation. Birkhäuser, Basel; Boston

72. Lew DH, Yoon JH, Hong JW, Tark KC (2010) Efficacy of antiadhesion barrier solution on periimplant capsule formation in a white rat model. Ann Plast Surg 65(2):254–258
73. Li J, Zhang YP, Kirsner RS (2003) Angiogenesis in wound repair: angiogenic growth factors and the extracellular matrix. Microsc Res Tech 60(1):107–114
74. Madlener M, Parks WC, Werner S (1998) Matrix metalloproteinases (MMPs) and their physiological inhibitors (TIMPs) are differentially expressed during excisional skin wound repair. Exp Cell Res 242(1):201–210
75. Maller O, Martinson H, Schedin P (2010) Extracellular matrix composition reveals complex and dynamic stromal-epithelial interactions in the mammary gland. J Mammary Gland Biol Neoplasia 15(3):301–318
76. Manuel JA, Gawronska-Kozak B (2006) Matrix metalloproteinase 9 (MMP-9) is upregulated during scarless wound healing in athymic nude mice. Matrix Biol 25(8):505–514
77. Matrisian LM, Wright J, Newell K, Witty JP (1994) Matrix-degrading metalloproteinases in tumor progression. Princess Takamatsu Symp 24:152–161
78. Metzmacher I, Ruth P, Abel M, Friess W (2007) In vitro binding of matrix metalloproteinase-2 (MMP-2) MMP-9 and bacterial collagenase on collagenous wound dressings. Wound Repair Regen 15(4):549–555
79. Morrison CJ, Overall CM (2006) TIMP independence of matrix metalloproteinase (MMP)-2 activation by membrane type 2 (MT2)-MMP is determined by contributions of both the MT2-MMP catalytic and hemopexin C domains. J Biol Chem 281(36):26528–26539
80. Muller M, Trocme C, Lardy B, Morel F, Halimi S, Benhamou PY (2008) Matrix metalloproteinases and diabetic foot ulcers: the ratio of MMP-1 to TIMP-1 is a predictor of wound healing. Diabet Med 25(4):419–426
81. Murshed M, McKee MD (2010) Molecular determinants of extracellular matrix mineralization in bone and blood vessels. Curr Opin Nephrol Hypertens 19(4):359–365
82. Nagase H, Woessner JF Jr (1999) Matrix metalloproteinases. J Biol Chem 274(31):21491–21494
83. Nakamura Y, Sato K, Wakimoto N, Kimura F, Okuyama A, Motoyoshi K (2001) A new matrix metalloproteinase inhibitor SI-27 induces apoptosis in several human myeloid leukemia cell lines and enhances sensitivity to TNF alpha-induced apoptosis. Leukemia 15(8):1217–1224
84. Neamtu M, Filioreanu AM, Petreus T, Badescu L, Ionescu CR, Cotrutz CE (2010) Involvement of MMP-8 in tissue response to colagenated fibrillar net in rats. Ann Roman Soc Cell Biol 15(1):236–241
85. Nilsen-Hamilton M, Werb Z, Keshet E (2003) Tissue remodeling. New York Academy of Sciences, New York, NY
86. Nomura H, Fujimoto N, Seiki M, Mai M, Okada Y (1996) Enhanced production of matrix metalloproteinases and activation of matrix metalloproteinase 2 (gelatinase A) in human gastric carcinomas. Int J Cancer 69(1):9–16
87. Overall CM, Kleifeld O (2006) Tumour microenvironment—opinion: validating matrix metalloproteinases as drug targets and anti-targets for cancer therapy. Nat Rev Cancer 6(3):227–239
88. Parks WC, Mecham RP (1998) Matrix metalloproteinases. Academic, San Diego
89. Peled ZM, Phelps ED, Updike DL, Chang J, Krummel TM, Howard EW et al (2002) Matrix metalloproteinases and the ontogeny of scarless repair: the other side of the wound healing balance. Plast Reconstr Surg 110(3):801–811
90. Pepper MS (2001) Role of the matrix metalloproteinase and plasminogen activator-plasmin systems in angiogenesis. Arterioscler Thromb Vasc Biol 21(7):1104–1117
91. Petreus T, Cotrutz CE, Neamtu M, Buruiana EC, Sirbu PD, Neamtu A (2010) Molecular docking on recomposed versus crystallographic structures of Zn-dependent enzymes and their natural inhibitors. Proc World Acad Sci Eng Technol 68:1992–1995
92. Petreus T, Cotrutz CE, Neamtu M, Buruiana EC, Sirbu PD, Neamtu A (2010) Understanding the dynamics–activity relationship in metalloproteases: ideas for new inhibition strategies. In: AT-EQUAL 2010: 2010 ECSIS Symposium on advanced technologies for enhanced quality of life: LAB-RS and ARTIPED 2010, pp 83–86

93. Ra HJ, Parks WC (2007) Control of matrix metalloproteinase catalytic activity. Matrix Biol 26(8):587–596
94. Rasmussen HS, McCann PP (1997) Matrix metalloproteinase inhibition as a novel anticancer strategy: a review with special focus on batimastat and marimastat. Pharmacol Ther 75(1):69–75
95. Rayment EA, Upton Z, Shooter GK (2008) Increased matrix metalloproteinase-9 (MMP-9) activity observed in chronic wound fluid is related to the clinical severity of the ulcer. Br J Dermatol 158(5):951–961
96. Reijerkerk A, Kooij G, van der Pol SM, Khazen S, Dijkstra CD, de Vries HE (2006) Diapedesis of monocytes is associated with MMP-mediated occludin disappearance in brain endothelial cells. FASEB J 20(14):2550–2552
97. Reiter LA, Freeman-Cook KD, Jones CS, Martinelli GJ, Antipas AS, Berliner MA et al (2006) Potent selective pyrimidinetrione-based inhibitors of MMP-13. Bioorg Med Chem Lett 16(22):5822–5826
98. Renault MA, Losordo DW (2007) The matrix revolutions: matrix metalloproteinase vasculogenesis and ischemic tissue repair. Circ Res 100(6):749–750
99. Rodriguez R, Loske AM, Fernandez F, Estevez M, Vargas S, Fernandez G et al (2010) In vivo evaluation of implant-host tissue interaction using morphology-controlled hydroxyapatite-based biomaterials. J Biomater Sci Polym Ed 22(13):1799–1810
100. Rosenblum G, Meroueh S, Toth M, Fisher JF, Fridman R, Mobashery S et al (2007) Molecular structures and dynamics of the stepwise activation mechanism of a matrix metalloproteinase zymogen: challenging the cysteine switch dogma. J Am Chem Soc 129(44):13566–13574
101. Ruifrok AC, Katz RL, Johnston DA (2003) Comparison of quantification of histochemical staining by hue-saturation-intensity (HSI) transformation and color-deconvolution. Appl Immunohistochem Mol Morphol 11(1):85–91
102. Rush TS 3rd, Powers R (2004) The application of x-ray NMR and molecular modeling in the design of MMP inhibitors. Curr Top Med Chem 4(12):1311–1327
103. Salmela MT, Pender SL, Karjalainen-Lindsberg ML, Puolakkainen P, Macdonald TT, Saarialho-Kere U (2004) Collagenase-1 (MMP-1) matrilysin-1 (MMP-7) and stromelysin-2 (MMP-10) are expressed by migrating enterocytes during intestinal wound healing. Scand J Gastroenterol 39(11):1095–1104
104. Salonurmi T, Parikka M, Kontusaari S, Pirila E, Munaut C, Salo T et al (2004) Overexpression of TIMP-1 under the MMP-9 promoter interferes with wound healing in transgenic mice. Cell Tissue Res 315(1):27–37
105. Santos MA, Marques S, Gil M, Tegoni M, Scozzafava A, Supuran CT (2003) Protease inhibitors: synthesis of bacterial collagenase and matrix metalloproteinase inhibitors incorporating succinyl hydroxamate and iminodiacetic acid hydroxamate moieties. J Enzyme Inhib Med Chem 18(3):233–242
106. Scatena R (2000) Prinomastat a hydroxamate-based matrix metalloproteinase inhibitor. A novel pharmacological approach for tissue remodelling-related diseases. Expert Opin Investig Drugs 9(9):2159–2165
107. Shekaran A, Garcia AJ (2011) Extracellular matrix-mimetic adhesive biomaterials for bone repair. J Biomed Mater Res A 96(1):261–272
108. Steward WP, Thomas AL (2000) Marimastat: the clinical development of a matrix metalloproteinase inhibitor. Expert Opin Investig Drugs 9(12):2913–2922
109. Stratmann B, Farr M, Tschesche H (2001) MMP-TIMP interaction depends on residue 2 in TIMP-4. FEBS Lett 507(3):285–287
110. Sung HJ, Johnson CE, Lessner SM, Magid R, Drury DN, Galis ZS (2005) Matrix metalloproteinase 9 facilitates collagen remodeling and angiogenesis for vascular constructs. Tissue Eng 11(1–2):267–276
111. Takahashi H, Akiba K, Noguchi T, Ohmura T, Takahashi R, Ezure Y et al (2000) Matrix metalloproteinase activity is enhanced during corneal wound repair in high glucose condition. Curr Eye Res 21(2):608–615

112. Tang L, Eaton JW (1995) Inflammatory responses to biomaterials. Am J Clin Pathol 103(4):466–471
113. Tang L, Hu W (2005) Molecular determinants of biocompatibility. Expert Rev Med Devices 2(4):493–500
114. Tanner KE (2010) Bioactive composites for bone tissue engineering. Proc Inst Mech Eng H 224(12):1359–1372
115. Terasaki K, Kanzaki T, Aoki T, Iwata K, Saiki I (2003) Effects of recombinant human tissue inhibitor of metalloproteinases-2 (rh-TIMP-2) on migration of epidermal keratinocytes in vitro and wound healing in vivo. J Dermatol 30(3):165–172
116. Thevenot P, Hu W, Tang L (2008) Surface chemistry influences implant biocompatibility. Curr Top Med Chem 8(4):270–280
117. Underwood CK, Min D, Lyons JG, Hambley TW (2003) The interaction of metal ions and Marimastat with matrix metalloproteinase 9. J Inorg Biochem 95(2–3):165–170
118. Vaalamo M, Leivo T, Saarialho-Kere U (1999) Differential expression of tissue inhibitors of metalloproteinases (TIMP-1 -2 -3 and -4) in normal and aberrant wound healing. Hum Pathol 30(7):795–802
119. Velasco G, Pendas AM, Fueyo A, Knauper V, Murphy G, Lopez-Otin C (1999) Cloning and characterization of human MMP-23 a new matrix metalloproteinase predominantly expressed in reproductive tissues and lacking conserved domains in other family members. J Biol Chem 274(8):4570–4576
120. Willenbrock F, Murphy G, Phillips IR, Brocklehurst K (1995) The second zinc atom in the matrix metalloproteinase catalytic domain is absent in the full-length enzymes: a possible role for the C-terminal domain. FEBS Lett 358(2):189–192
121. Woessner JF Jr (1994) The family of matrix metalloproteinases. Ann N Y Acad Sci 732:11–21
122. Woessner JF Jr (2001) That impish TIMP: the tissue inhibitor of metalloproteinases-3. J Clin Invest 108(6):799–800
123. Wolber G, Langer T (2005) LigandScout: 3-D pharmacophores derived from protein-bound ligands and their use as virtual screening filters. J Chem Inf Model 45(1):160–169
124. Wolber G, Seidel T, Bendix F, Langer T (2008) Molecule-pharmacophore superpositioning and pattern matching in computational drug design. Drug Discov Today 13(1–2):23–29
125. Zhang W, Hou TJ, Qiao XB, Huai S, Xu XJ (2004) Binding affinity of hydroxamate inhibitors of matrix metalloproteinase-2. J Mol Model 10(2):112–120
126. Zohny SF, Fayed ST (2010) Clinical utility of circulating matrix metalloproteinase-7 (MMP-7) CC chemokine ligand 18 (CCL18) and CC chemokine ligand 11 (CCL11) as markers for diagnosis of epithelial ovarian cancer. Med Oncol 27(4):1246–1253
127. Zucker S, Drews M, Conner C, Foda HD, DeClerck YA, Langley KE et al (1998) Tissue inhibitor of metalloproteinase-2 (TIMP-2) binds to the catalytic domain of the cell surface receptor membrane type 1-matrix metalloproteinase 1 (MT1-MMP). J Biol Chem 273(2):1216–1222

Chapter 3
Synthetic Morphogens and Pro-morphogens for Aided Tissue Regeneration

Matteo Santin

3.1 Introduction: Morphogen Definition

The main goal of developmental biology has been the understanding of tissue morphology that is the mechanism by which tissue and organ structures form during development. At cellular level, the fundamental question to be answered is how cellular spacing and positioning within a tissue are controlled to form histological and anatomical structures. It is now accepted that in a forming tissue there is a specific group of cells, called "*organisers,*" that acquires a role of instructing neighbour cells to exert specific functions [1]. At molecular level, these instructions are biochemical signals, mainly proteins, which are received and interpreted by the neighbour cells. This interpretation is not processed by the neighbour cells through an "*on/off-mode*" mechanism, but it is modulated by a molecular machinery that is able to quantitatively measure the strength and the distance of the signal, thus assessing the distance of the neighbour cell from a given position. Such a type of biochemical signal is called "*morphogen*" (Fig. 3.1) [2].

The term morphogens was coined by the famous mathematician Alfred Turing before the introduction of the concept of biochemical signalling [1]. Indeed, the concept that some kind of cellular signalling could be present in developing organisms was first introduced by two embryologists, Dalq and Pasteels, who suggested

M. Santin (✉)
Brighton Studies in Tissue-Mimicry and Aided Regeneration (BrightSTAR),
School of Pharmacy and Biomolecular Sciences, University of Brighton,
Huxley Building Lewes Road, Brighton BN2 4GJ, UK
e-mail: m.santin@brighton.ac.uk

I. Antoniac (ed.), *Biologically Responsive Biomaterials for Tissue Engineering*,
Springer Series in Biomaterials Science and Engineering 1,
DOI 10.1007/978-1-4614-4328-5_3, © Springer Science+Business Media New York 2013

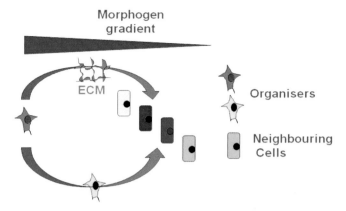

Fig. 3.1 Schematic representation of morphogen-based signalling in developing organisms. An organiser generates a morphogen gradient that reaches neighbour cells. The neighbour cells differently interpret the signal depending on their position in the gradient. Intermediate cells (*lower pathway*) and macromolecules of the extracellular matrix (*upper pathway*) contribute to the maintaining and control of the gradient

that, during the development of an embryo, signals could diffuse from cell to cell to organise tissues.

According to the current opinion, for a molecular signal to be recognised as a morphogen two main criteria need to be fulfilled [1]:

- It should be able to act directly at a distance.
- It has to induce distinct biological/morphological responses at different concentrations.

Most studies on morphogens have therefore been performed on developing organsims and, in particular, on invertebrates (e.g. the fruit fly *Drosophila melanogaster*) [3]. Here, morphogens that determine the tissue patterning (e.g. the anterior–posterior axis of the body) have been the first to be identified and include transcriptional factors able to activate specific genes in the cells such as the Hedgehog (Hh) and the decapentaplegic (dpp) factors.

Studies have shown the ability of Hh to act both directly at a short distance or to activate dpp that is a long-distance factor of tissue patterning [1].

However, it is also important to highlight that the ability of the morphogens to activate specific genes in the target cells also depends on the expression of specific proteins in the latter [1, 4].

3.2 Morphogens in Vertebrates

The proof of the existence of morphogens in vertebrates has not been as straightforward as in invertebrates [5]. While studying the development of limbs, a morphogen gradient able to control their position in relation to a group of mesenchymal cells posterior to the embryo has been proposed. This gradient has been named the zone of polarising activity (ZPA), the source of the signal that controls the pattern of digits [2]. Although retinoic acid was early identified as the potential candidate for this signalling activity, later secondary signals emerged which were identified as the bone morphogenetic protein-2 (BMP2) that is the homologue of the invertebrate dpp and the Sonic Hedgehog (Shh) that is the homologue of the invertebrate Hh [2]. However, it has to be said that it is not yet clear if these factors really satisfy the criteria established for the definition of morphogens (See Sect. 3.1).

3.3 Morphogen Long-Distance Transport and Local Mechanisms

The need for morphogens to exert their function in a gradient-dependent fashion has prompted a number of studies about the molecular basis generating these gradients [1, 3, 6]. In particular, it is not clear how the diffusion of morphogens within tissues is regulated. Developing tissues such as the epithelia can present a relatively complex structure preventing an easy diffusion of macromolecules [1]. It has been suggested that cell proliferation can contribute to long-distance targeting of specific signals and that these signals can be internalised by intermediate cells by endocytosis to be later released in the extracellular space (Fig. 3.1, lower pathway) [1]. This process is known as planar transcytosis. In addition, it has been suggested that signal may diffuse in the extracellular space not as soluble molecules, but forming complexes with cell surface macromolecules such as the heparan sulphate proteoglycans (Fig. 3.1, upper pathway) [1]. These are also claimed to directly influence various aspects of the morphogen gradients including their movement, signalling and trafficking [7]. A mathematical model for morphogen gradient formation suggests that morphogens are synthesised in local regions to form long-distance gradients through the binding to heparan sulphate proteoglycans chains thus reaching target cells.

It is also known that molecular gradients within a tissue can take place without the occurrence of molecular diffusion in the extracellular space [8]. It has been suggested that cell-mediated degradation of morphogens can generate gradients regardless of their ability to diffuse in the extracellular space. Transport from cell to cell through gap junctions can also occur and cells could be the transport vehicle of the molecule. These kinds of transport might enable the formation of gradients of non-secreted molecules. Molecular gradients can also be formed by the dilution of the molecular content of cells that continuously divide and become displaced away from a source.

It has been highlighted that for a morphogen gradient to effectively work, precision and robustness of the signalling are necessary [1].

The level of precision that is required has not yet been clarified, but in the Drosophila embryo there are evidences of signalling precision both at cellular and nuclear level [9].

A second condition to be fulfilled to generate spatial patterns during embryogenesis is robustness. This means that the distribution of morphogens must be reproducible and provide a reliable signalling regardless of the occurrence of other processes that can alter their distribution [10]. For example, the same morphogen binding sites of the extracellular matrix (ECM) can be altered, thus becoming a hindrance to the regular diffusion of these molecules [11]. It has been calculated that the complex geometry of the ECM can increase the diffusion times of morphogens as much as fivefold [12]; this tortuous space also contributes to the generation of local variations in the morphogen gradient. In summary, morphogen diffusion will depend on the three-dimensional structure of the tissue as well as on the density and complex topology of cell-to-cell contacts. Indeed, soluble morphogens can bind to cell-surface receptors and be internalised by a process called endocytosis [1].

The interaction between morphogens and cells is not only limited to the biospecific recognition mediated by receptors. The need for morphogens to play a role as positional cues requires a mechanism in the cell that enables the interpretation of a given concentration within the gradient (Fig. 3.1). If the position of a cell within a tissue is determined by signals at single-cell level, it can be speculated that morphogen concentration thresholds need to be present so that different genes can be activated at given morphogen concentration levels. The binding of morphogens to receptors can also play a role on morphogen distribution as the saturation of the receptors can block the gradient effect [13].

Wolpert et al. [14] have proposed a model of morphogen gradient response by the cells in the embryo. In this model, there would be an initial state where all the cells in a given region are equivalent, but they are all able to enter several phenotypes. These phenotypes are acquired depending on the space distribution of the morphogens. With time, the morphogen distribution becomes stable and at this stage each cell reads the concentration and compares it with a set of threshold to finally respond only to the signal that is present at the cell location. However, it has to be outlined that the effect of time on the effectiveness of the morphogen gradient has not yet been proven.

Other alternative mechanisms and structures that regulate morphogen gradients have been suggested. For example, during Drosophila development, cytonemes which are actin-based filopodial extensions appear to be oriented preferentially towards both the anteroposterior and dorsoventral organiser cells and therefore it has been suggested that they can be involved in the DPP signalling mechanism [1].

Together with cell-to-cell interactions, morphogens have also been indicated as possible regulators of cell polarity. Indeed, for a tissue to form it is important that cells can assume polarity, i.e. the ability to acquire directional information. This property is important for the final structure of the tissue and consequently for its functionality.

It has to be said that it appears improbable that morphogen concentration is the only factor determining cell fate within the developing embryo. Therefore, morphogens may only be part of the signalling dictating cell fate and position. Cell position could be more accurately tuned by cell–cell interactions.

3.4 Growth Factors as Morphogens

In 1991, Green and Smith [15] proposed growth factors as alternative candidates for natural morphogens. As the morphogens, growth factors are also known to regulate cell-to-cell communication. These authors focussed on the role of growth factors in embryo development and highlighted that mesoderm-inducing factors (MIFs, e.g. the activin) are related to the fibroblast growth factor (FGF) and to the superfamily of the transforming growth factor beta (TGF-s). In Sect. 3.2 of this chapter, it has been mentioned that dpp is related to another important growth factor, the BMP-2, that is a member of the TGF-a superfamily. There are some experimental evidences suggesting that the Xenopus activin mesoderm-inducing factor (XTC-MIF) can yield notochord and neural tissue at high doses, while low doses induce the formation of both a mesenchyme and blood-like cells. Likewise, FGF has been shown to produce dose-dependent effects but the range of responses by the cells that it can induce appears to be more limited. TGFT-like and FGF act antagonistically to specify the precursor tissue leading to the formation of both the retina pigmented epithelium and the neural retina, whereas it has been observed that opposed increasing and decreasing gradients of BMP and Shh control the establishment of dorsoventral polarity of the optic cup [16]. Within this developmental process of the eye tissue, additional cell phenotype formation is controlled by morphogens/growth factors including Wnt, Shh, BMP and FGF.

Likewise, the process of axon wiring during nerve formation takes place through two main steps: path-finding and target selection [17]. During path-finding, neural cells extend from their main body growth cones which navigate towards their targets establishing a network of axons. When the axons reach the target area, they start to select their targets within a group of similar cells and establish the so-called synaptic connections that are very specifically organised to provide complex functionalities. As for the eye tissue formation, BMPs, Wnt and Shh are key bio-cues for the nerve developing process.

Many papers have shown the role of BMPs in pattern formation of tissues of the musculoskeletal system [5]. In particular, the role of BMPs in bone formation is well known. However, experiments of homologous recombination targeting the function of particular genes have demonstrated the actions of BMPs beyond bone in such disparate tissues as kidney, eye, testis, teeth, skin and heart [5].

3.5 Knowledge in Developmental Biology: Lessons for Therapeutically Driven Tissue Regeneration

Despite its complex and relatively unknown mechanisms, the pattern of tissue formation in embryogenesis offers a large knowledge platform that can be exploited for new clinical applications focussing on the regeneration of tissues damaged by either trauma or disease. In particular, despite the complexity of their mechanism(s) of action, morphogens stem out as of key importance in the signalling process that is required for tissue formation. Two main aspects of the morphogen role are worth noting:

- Their number appears to be relatively limited when compared to the number of functions they can exert on the developing embryo. From the examples provided in Sect. 3.4, it is obvious that a variety of tissues can be generated by only a few types of morphogens/growth factors. Indeed, it is significant that some of them are structurally and functionally related to growth factors controlling the tissue signalling pathways in adult organisms.
- Their interactions with the macromolecules of the ECM appear to be important in the regulation of the tissue formation. In particular, docking sites are necessary to control the diffusion and the local concentrations of the morphogens that are indicated as the main mechanisms for cell interpretation of the biosignalling.
- Their recognition by the cells appears to be a regulator not only of the cell function, but also of their gradients and local concentrations.
- Their actions through concentration gradients, opposing gradients and cell biorecognition are mainly exerted on stem cells presenting a totipotent or multipotent phenotype that allows them to differentiate in various types of tissue cells.

Most of these controlling mechanisms can be regarded as lessons to be transferred to the study of tissue repair and regeneration in adult organisms and to clinical applications aiming at the repair of tissues damaged by either trauma or disease.

3.6   Growth Factors in Adult Tissue Repair and Regeneration

Unlike embryo development, adult human morphogenesis during the process of tissue repair is guided by a variety of signals produced by different types of cells (e.g. inflammatory cells and tissue cells). Indeed, following damage, tissue repair is promoted by cytokines and growth factors that interact with tissue cells to regulate their genes and in turn their metabolic activity [18]. Traumatic events (accidental or surgical) and diseases cause the loss of the tissue architectural, biochemical, cellular and homeostatic properties [19]. Indeed, the ECM is mainly composed of structural proteins (e.g. collagens, fibronectin) and polysaccharides (e.g. glycosaminoglycans) which preserve cell viability and the delicate homeostatic balance of the tissue (Fig. 3.2a).

As discussed in Sect. 3.3, any change in the ECM structure can alter the distribution of the morphogens and therefore their action. As a consequence of its disruption it is likely that the biosignalling during healing may negatively impact the ability of the ECM macromolecules to control the morphogen/growth factor gradients.

The restoration of these properties in spontaneous healing is gradual and takes place through a number of phases starting from the arrest of the bleeding by the clot which is gradually digested by inflammatory cells and proteolytic enzymes to give

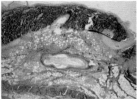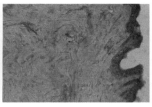

Fig. 3.2 Typical histological features of (**a**) healthy skin, (**b**) granulation tissue, (**c**) scar tissue. Oriented collagen fibres are the main feature of the healthy skin. The granulation tissue is characterised by a high density of small blood vessels, the cross section of which appears as "granules" throughout the tissue. The scar tissue shows collagen fibres highly packed and disordered when compared to the healthy tissue

space to new tissue [20]. The secretion of growth factors by inflammatory cells, the macrophages, stimulates the formation of a highly vascularised granulation tissue generally progressing into scar (Fig. 3.2b, c).

With time, the scar, a tissue characterised by highly packed collagen fibrils, undergoes remodelling into a tissue that only in some cases and with much delay resembles the features of the original healthy tissue. The biochemical stimuli secreted during the healing process recruit mesenchymal cells to the site of injury to induce not only their proliferation, but also their spatial organisation.

For example, in the case of skin repair, epithelial cell migration and proliferation are controlled to cover the temporary matrix of the underlying granulation tissue. Later, in the tissue repair process of the skin the underlying repairing dermis hosts the mutual stimulation of cells such as keratinocytes and fibroblasts. During the process, fibroblasts acquire a proliferative phenotype that not only facilitates the site colonisation but also produce the ECM contraction that facilitates wound closure. All these activities depend on a tuned balance between the secretion of pro-inflammatory cytokines and that of growth factors including TGF-alpha; here the link with embryogenesis becomes evident.

However, spontaneous tissue replace can be achieved only when some conditions are met:

- The area of damage still includes a macromolecular structure providing a scaffolding role for migrating and proliferating cells.
- The biochemical signalling is not interrupted by excessive distance.
- The cell activity is not compromised by any pathological unbalance.

In those cases where the area of damage reaches a critical size or alterations of the biochemical and cellular pathways are caused by pathological conditions (e.g. diabetes, autoimmune diseases) cell activities including their crosstalk are no longer able to establish active concentration gradients [20].

In the case of disease, precise alterations of the modality of interaction between growth factors and relative cell receptors have been identified. For example, studies by a diabetic mouse model have shown the relatively lower expression of receptors for the platelet-derived growth factor (PDGF-BB), an important stimulator of fibroblast proliferation and promoter of tissue repair [21]. In such a case, the presence of the growth factor as well as of its diffusion through a gradient may not be sufficient to ensure tissue repair.

3.6.1   Use of Growth Factors in Clinical Intervention

The use of growth factors as therapeutic tools has widely been explored at research level and, in some cases, it has reached its clinical use. Relevant to this chapter is the use of growth factors structurally and functionally related to embryo morphogens.

The most important class of growth factors used in bone tissue engineering is the TGF-m superfamily that includes the TGF-1 and the BMPs [22]. In adults, BMPs act as a stimulating factor for stem cell differentiation into osteoblasts and direct osteochondral calcification. BMP-2, also known as OP-1, has been commercialised by Stryker and used to treat bone critical defects. Conversely, FGF-2 has been investigated showing conflicting evidences about its bone repair potential.

Alongside these growth factors with morphogen similarities, other growth factors are worth mentioning because of their therapeutic potential for tissue repair. The insulin-like growth factor (IGF) is another growth factor that has been carefully investigated in relation to cartilage and bone repair [22].

Together with FGF, the vascular endothelial growth factor (VEGF) is widely recognised as the principal player in angiogenesis and it has been used to improve vascularisation of newly formed bone. In addition, it has been demonstrated that VEGF is also a mediator of various osteoinductive factors such as TGF-β1, thus providing an indirect stimulus to osteogenesis [22]. As a potent stimulator of fibroblast proliferation PDGF-BB has also been proposed as a growth factor in skin regeneration. The clinical performance of acellular skin grafts such as Biobrane®, Biobrane-L® and Transcyte®, bioengineered dermal substitutes, is linked to their ability to provide growth factors including PDGF-BB [23, 24].

Neural growth factor (NGF) is the target of many studies aiming at peripheral nerve regeneration. NGF is released by the distal end of a damaged nerve to attract axonal regeneration at the proximal stump. However, only short nerve gaps are capable of spontaneous repair. In an attempt to address this clinical problem, many research groups have studied the delivery of NGF into the lumen of a repairing nerve. One specific example of such an approach is the delivery of NGF into a silicone nerve guide by an osmotic pump [25]. This method of delivery has been proven to be successful in improving the quality of nerve repair over a 5 mm gap. However, it is important to consider that these experiments were performed in a healing model, thus not proving the clinical potential of this approach. Indeed, the delivery of NGF into the lumen of nerve guide has been evaluated in many publications. However, not all have been successful in improving the quality of nerve repair. Some success was reported in the publications by Piquilloud et al. [26] and Barras et al. [27] showing significant evidence in support of the NGF effect.

3.6.2 Limitation in the Use of Growth Factors in Clinical Intervention

In therapeutic applications, the use of growth factors is limited by their purification costs, batch-to-batch reproducibility, known instability under storage conditions and short lifespan in physiological conditions. In addition, their ability to stimulate cells tightly depends on their bioavailability. In most of the cases, short lifespan and sufficient bioavailability require the administration of relatively high dose of these growth factors. Such doses are not only very expensive, but they also increase the risks of tumours. Safety concerns are also caused by the risk of transmittable diseases from donor to recipient. While the costs and risks of transmittable diseases have been tackled by the synthesis of recombinant proteins, problems still remain about the stability under storage and administration conditions as well as about the relatively high dosage.

It is thought that by improving the growth factor stability in vivo, relatively lower doses will become therapeutically efficacious. Delivery systems which are able to preserve the native conformation of these growth factors and to provide appropriate concentrations and gradients in the extracellular space have been designed for this purpose. For example, growth factors for bone repair have been delivered through the use of polymeric, ceramics and composite biomaterials. The review papers by Lee and Shin [22] and by Seeherman and Wozney [28] offer an overview of the various combinations of biomaterials and growth factors employed for this purpose. In all these methods the slow delivery of these growth factors relies upon the physico-chemical properties of the carriers which are far to mimic the docking sites for morphogens and growth factors present in the ECM of both embryonic and adult tissues.

FGF-1 localised delivery has also been pursued through the entrapment of the growth factor in a fibrin/hydroxyapatite composite. In vivo this system was capable of delivering biologically active FGF-1 for a longer period of time leading both to enhanced angiogenesis and infiltration of osteoprogenitor cells [29].

Similarly, FGF-2 release from p(HEMA-co-VP), while showing that FGF-2 was able to enhance bone formation by increasing bone volume, did not show a protracted effect after surgery [30].

The methods of NGF delivery proposed by Piquillod et al. [26] rely on a degradation-linked release effect from the polymeric matrix, while the approach followed by Barras et al. [27] capitalises on the diffusion of the entrapped growth factor from the neural guide mesh. As outlined above, these methods of delivering present significant limitations as, for example, degradation-driven release may limit the growth factor bioavailability, while in the case of simple entrapment the half-life of NGF may be the principal limiting factor.

Indeed, as nature has developed fine interactions of reversible binding between the growth factors and the ECM components morphogens and growth factors are bound to them until they are locally mobilised by cells [11]. This has been the rationale to develop specific biomaterials able to dock growth factors on their surface.

Fibrin-based biomaterials for bone tissue engineering have been developed that present a BMP-2 fusion protein which is released under the action of proteases associated to the cell surface [31]. The protein has been engineered to present an N-terminal transglutaminase substrate domain providing covalent attachment to fibrin during its polymerisation. A central plasmin substrate domain provides a cleavage site for proteases localising the release of the attached growth factor. The enhanced bone repair process obtained by these engineered biopolymers seems to corroborate the view that morphogen and growth factor gradients and localised concentrations are necessary for tissue repair.

Likewise, the reversible binding of BMP-2 has been pursued by conjugating heparin to a synthetic poly(L-lactic-co-glycolic acid) (HP-PLGA) scaffold [32]. This conjugation was enhanced by the use of star-shaped polymers sustaining the release for over 14 days in vitro and the osteoblast differentiation in vitro over 14 days was significantly higher than non-heparinised PLGA. In vivo bone formation by the BMP-2-loaded HP-PLGA scaffolds was fourfold greater than BMP-2-loaded unmodified PLGA scaffold group.

3.7 Synthetic Morphogens, Growth Factors and Bioligands

Morphogens, growth factors and cell activity regulators are mostly proteins or relatively short peptides. These native molecules are relatively large, but several studies have identified that specific amino acid sequences within their primary structure are responsible for mediating biointeractions and bioactivity of cells. These findings have paved the way towards the production of completely synthetic growth factor analogues and bioligands. Indeed, their use in clinical applications of tissue repair and regeneration could bring the following advantages when compared to growth factors of animal, human or recombinant origin:

- Reduced synthesis and purification costs.
- Elimination of safety risks associated to contamination of the product by transmittable pathogens.
- High batch-to-batch reproducibility.
- Tailored design for conjugation of the bioactive sequence to biomaterial surfaces. This would allow the docking of the peptide to biomaterial scaffolds used for tissue repair or regeneration.

3.7.1 Identification of Bioactive Sequences by Phage Displayed Peptide Technique

Phage displayed peptide technique has been the main technological tool to identify peptide sequences able to recognise cell receptors [33]. Phage displayed peptide technology is based on the ability of bacteriophage to express foreign (poly)peptides as fused sequences to the proteins of their capsid surface. This technique was described for the first time by George P. Smith [34]. The display of specific peptides on the bacteriophage capsid surface is obtained by encoding a gene for a specific peptide into the gene for a capsid structural protein. As a result the phage will synthesise billions of pooled peptides which will be presented on phage particles to constitute a phage-displayed peptide library. Unlike other methods, as many as 1,010 different peptides can be screened simultaneously for a targeted activity [35]. The selection of peptides from these phage libraries focuses on the interaction with proteins and enzymes often to interfere with their activity [36]. Since its discovery, this technique has been used to design new agonists and antagonists for cell receptors [37], for the development of vaccines and for the selection of new antibodies [38, 39] as well as for improving the analysis of protein-to-protein interaction in proteomics [40] and to identify new substrates and inhibitors in enzymology [41]. Phage-displayed peptide libraries have been obtained by either lytic or filamentous phages or by phagemid vectors [42].

In the case of targeting of cell receptors, biopanning is considered the method of choice as, from the initial bacteriophage pool, it provides small numbers of clones, each one of them being specific for a single peptide and offering the desired properties in terms of affinity or activity [33]. To select the peptide affinity the phages are introduced to an immobilised target, the unbound phages are then removed by washing and a final elution step allows the removal of the bound phages. Typically, to ensure the selection of the high-affinity clones a few washing and elution steps are performed. Once eluted, the specific phages are amplified in host bacteria and undergo cyclical selection steps under more defined conditions that significantly enhance their specificity [35]. In some cases, a negative selection step can also be performed to increase the efficiency of the selection process. Once the selection process has been completed, the primary structure of the displayed peptide is easily identified by determining the nucleotide sequence of the corresponding insert [35].

3.7.1.1 Identification of Growth Factor Analogues

Recently, phage displayed peptide technique has allowed the identification of the structure and function of a number of cytokines and growth factors and their receptors [43].

The restricted sequences of the OP-1 and several sequences of the VEGF have been identified alongside with VEGF blockers and VEGF receptors 1 and 2 antagonists. In addition, sequences mimicking the PDGF-BB, NGF and IGF-1 have been

Table 3.1 Synthetic peptidic growth factor analogues (adapted from Santin M, [46])

Peptide sequence/name	Growth factor analogue/function	Reference
SSRS	Insulin growth factor-1 analogue/ corneal regeneration	Yamada et al. [44]
RKIEIVRKK	Platelet-derived growth factor-BB analogue/angiogenesis	Brennand et al. [45]
QPWLEQAYYSTF YPHIDSLGHWRR LLADTTHHRPWT AHGTSTGVPWP VPWMEPAYQRFL TLPWLEESYWRP	Vascular endothelial growth factor analogues/angiogenesis	Hardy et al.[46]
KLTWQELYQLKYKGI	Vascular endothelial growth factor analogue/angiogenesis	D'Andrea et al. [47]
NSVNSKIPKACCVPTELSAI	Bone morphogenetic protein-2 analogue/bone regeneration	Suzuki et al. [48]
–GGSKPPGTSS–CONH$_2$	Bone marrow homing/cell differentiation	Horii et al. [49]
–GGPPSSTKT–CONH$_2$	Bone marrow homing/cell differentiation	Horii et al. [49]

identified. Table 3.1 provides a list of peptides with a growth factor analogue activity and their relative literature references.

In addition to these analogues, peptides which are involved in vivo in the homing of bone marrow and haemoprogenitor stem cells have also been identified. Selected peptides also include amino acid sequences with an angiopoietin-1 mimicking sequence, NQRRNPENGGRRYNRIQHGQ, or the SFLLRN sequence that is able to commit haemoprogenitor cells towards angiogenesis through induction of angiopoietin-2 expression. Indeed, both angiopoietin-1 and -2 play an important biological role in angiogenesis.

Angiopoietin-1 is known to have both anti-inflammatory and angiogenic effect [50], while angiopoietin-2 induces PAR-1-mediated proliferation of haemoprogenitor cells [51].

Particularly relevant to tissue regeneration is also a class of peptides with a sequence able to recruit mesenchymal stem cells. It is envisaged that gradients of this peptide can enhance the regeneration potential of tissues compromised by trauma or disease.

3.7.1.2 Identification of Bioligand Analogues

Likewise for growth factor analogues, a number of peptides have been identified that are able to recognise cell receptors involved in cell biospecific recognition during the adhesion and migration processes. Table 3.2 provides a list of these peptides and their relative literature references. It can be observed that alongside the well-known RGD sequence that mediates the adhesion of a number of tissue cells and of

Table 3.2 Adhesion peptide sequences used in biomimetic materials (adapted from Pollock and Healy, 2009) [52]

Sequence	Analogue protein	Targeted cell receptor	Reference
RGD(X) (and di-cysteine cyclic forms)	Collagen (X=T), fibronectin (X=S), vitronectin (X=V), laminin (X=N), thrombospondin, osteopontin, bone sialoprotein	$\alpha v \beta 3$, $\alpha 5 \beta 1$, $\alpha 3 \beta 1$ and other integrins	Hern and Hubbell [53], Schense and Hubbell [54], Shin et al. [55], Harbers and Healy [56], Brandley and Schnaar [57]
DVDVPDGRGDLAYG	Osteopontin	$\alpha 5 \beta 1$ integrin	Shin et al. [58]
CGGNGEPRGDTYRAY	Bone sialoprotein	αv and $\beta 1$ integrins	Harbers and Healy [56]
FHRRIKA	Bone sialoprotein	Transmembrane proteoglycans	Rezania and Healy [59]
DGEA	Collagen-I	$\alpha 2 \beta 1$ Integrin	Schense and Hubbell [54]
IKVAV	Laminin-α	$\alpha 1 \beta 1$ and $\alpha 3 \beta 1$ Integrins	Silva et al. [60]
YIGSR	Laminin-β-1	110 and 67 kDa Laminin-binding protein	Schense et al. [61]

inflammatory cells (i.e. monocytes/macrophages), other sequences have been identified that are able to support osteoblast migration (i.e. FHRRIKA).

3.8   Current Limitations and Future Strategies to Improve the Therapeutic Potential of Analogue Morphogens, Growth Factors and Bioligands

The use of analogue morphogens, growth factors and bioligands in regenerative medicine is certainly advocated as they allow a highly reproducible production of biologically active molecules with no risk of transmittable disease. However, although still in their infancy, limitations are already emerging that may prevent their use in combination with biomaterials. It is therefore important to critically assess these limitations in the view of the final clinical applications and to draw new strategies aiming at improving the therapeutic potential of these peptides.

3.8.1   Current Limitations

One of the main possible limitations in the use of synthetic morphogens, growth factors and bioligands in regenerative medicine is the binding affinity of the isolated peptides that in other therapeutic applications has been shown to be relatively low [62]. However, it has been demonstrated that the efficacy of these peptides can be improved when conjugated with other molecules and carrier surfaces [63]. In the case of bioligands, this is not a significant problem as they are designed to be grafted on the surface of biomaterials to support cell adhesion and migration. Likewise, the half-life of these relatively small peptides when administered in vivo is also short as they are prone to degradation by hydrolysis and immune response. As discussed for the natural growth factors and due to the potential low affinity for the cell receptors, the bioavailability can also represent a significant limitation in the therapeutic use of these molecules. From an industrial viewpoint, it is also worth mentioning that the production (i.e. synthesis and purification) of these peptides can also be limited by the feasibility of their scaling up. Addressing this issue appears to be the most relevant step forward in the attempt to translate research findings into tools for clinical treatments.

3.8.2   Future Strategies

Inevitably, future strategies to improve the therapeutic potential of analogue morphogens, growth factors and bioligands have to mirror their current limitations and be oriented towards molecular and biomaterial designs specific for their integration into specific clinical applications and their relative procedures.

3.8.2.1 Solid-Phase Peptide Synthesis

Peptide synthesis methods have provided an opportunity to synthesise highly reproducible and scalable batches of peptides with analogue bioactivity of morphogens, growth factors and bioligands. In particular, the solid-phase synthesis approach appears particularly attractive as it allows a fine control of the product quality and the simplification of purification procedures [64]. Indeed, the scaled-up synthesis of therapeutically active peptides has already been applied to the pharmaceutical field and its application to the production of medical device and regenerative medicine products is therefore realistic. There are various strategies to synthesise peptides by solid-phase synthesis. The common denominator of these strategies is a process of gradual addition of amino acid units to a resin previously functionalised with a linker. Resins are available on the market that expose functional groups such as $-NH_2$ and $-COOH$ that are able to anchor molecules through peptidic bonds. These resins are usually functionalised with linker molecules that are susceptible to cleavage by specific organic solvents. The presence of this cleavable linker allows the liberation of the final peptide product from the resin. The peptide product is obtained through the addition of amino acids according to the sequence identified as bioactive by phage display technique. To ensure control over this process, amino acid with protecting groups has been made available on the market. Different types of protecting groups (e.g. Fmoc or Bmoc groups) have been identified to facilitate various conditions of synthesis. These functional groups can block either the $-NH_2$ or $-COOH$ of the amino acid. Once grafted to the linker or to the previously attached amino acid, the protecting group is cleaved off from the amino acid to expose the functional group (i.e. $-NH_2$ or $-COOH$) necessary for the formation of the following peptidic bond. Specific reagents activate the exposed functional groups to allow this reaction. Hence, the synthesis of the peptide can proceed towards it carboxy-terminal (C-terminal) or amino-terminal (N-terminal) end. Table 3.3 summarises the various steps of a typical protocol of solid-phase synthesis as applied to the synthesis of a BMP-2 analogue.

3.8.2.2 Hyperbranched Molecular Scaffolds

The concept of protein scaffolds is distinct from biomaterial scaffolds. Protein scaffolds are generally understood to encompass a range of protein domain-based frameworks that are neither conventional antibodies nor peptides [65]. These molecular scaffolds are used to stabilise the conformation of bioactive peptides and to maximise their exposure to the biochemical or cellular component with which they are destined to interact. In other words, protein scaffolds aim at replacing or mimicking the tertiary structure of proteins that determine the molecular folding

Table 3.3 Typical protocol for the solid-phase synthesis and characterisation of a growth factor analogue (relevant amino acid sequences are reported in Table 3.1)

Growth factor mimic synthesis
- Peptides were synthesised by a solid-phase synthesis method on a microwave system (Biotage) at mg scale
- Amino acids were loaded sequentially into the reaction vessel and coupled to 0.5 g resin (0.1 mmol $-NH_2$) as follows:
  - Rink Amide linker and HBTU/DIPEA
  - Fmoc-amino acid –OH and HBTU/DIPEA
  - Sequential addition of ×4 molar excess of each amino acid (0.4 mmol) in turn
- Deprotection step 20 % piperidine in DMF was added to remove Fmoc groups
- Peptides cleaved from resin using Trifluoroacetic acid:H_2O:Triisopropyl silane (95:2.5:2.5) and precipitated in cold diethyl ether. A series of centrifugation and diethyl ether washing steps are used to collect peptide
- Peptides were purified by preparative HPLC
- Peptides were characterised by
  - Analytical HPLC
  - Mass spectrometry (microTOF)
  - Gravimetric analysis

and the native conformation. Protein scaffolds have been engineered through different biotechnological approaches and include [66]:

- Cystine knot miniproteins derived from plant cyclotides able to present bioactive peptides
- Tetracorticopeptide modules presenting a variable paratope based on a natural protein fold
- Human I-set domains encompassing cell adhesion molecules

Peptide-based dendrimers have been proposed as scaffolds capable of enhancing epitope presentations and of mimicking protein secondary structures.

Dendrimers are highly and 3D ordered, hyperbranched polymers forming nanostructures with tuneable physico-chemical properties. Dendrimers are obtained from a core molecule to which synthetic (e.g. polyamido amine, PAMAM), amino acid (e.g. lysine) and carbohydrate monomers are bonded [67]. There are two main methods to synthesise dendrimers [68]:

1. The divergent synthesis where a core molecule with multiple reactive sites is used to form chemical bonding with a reactant and where the formed complex is later reacted with a molecule capable of generating another branching point
2. The convergent synthesis where fragments of dendrimers are added to the core molecules and thus assembled

When the synthesis is performed in liquid phase, although the shape and symmetry of the dendrimer depend on the physico-chemical properties of the molecules used for its synthesis, the polymer branching generally leads to an open ball, spherical structure [69] (Fig. 3.3).

Fig. 3.3 Schematic representation of a dendrimer obtained by liquid-phase synthesis. Gn indicates the number of branching generations (Figure kindly provided by Dr. Steven Meikle, University of Brighton)

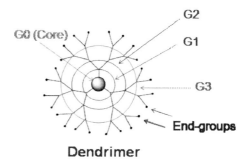

Dendrimer

Fig. 3.4 Schematic representation of a semi-dendrimer (dendron) obtained by a solid-phase method. Gn indicates the number of branching generations (Figure kindly provided by Dr. Steven Meikle, University of Brighton)

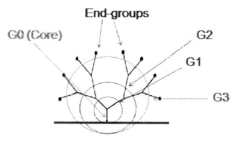

Semi-dendrimer or Dendron

Conversely, when the synthesis is performed on a solid phase the branching polymer develops a dome-like (semi-sphere) or a tree-like structure, the semi-dendrimer (or dendron) [70] (Fig. 3.4).

By both methods it is possible to obtain dendrimers (or dendron) with several branching levels (generations, Gn) (Figs. 3.3 and 3.4). The synthesis of dendrimers up to G9 has been reported [71]. From a biotechnological viewpoint both dendrimers and dendrons offer a unique opportunity to expose functionalities able to favour bio-interactions by a nanostructurally controlled manner. Dendrimers have been mainly proposed as carriers for the delivery of contrast agents to tissues as well as of nucleic acid drugs to cells (e.g. US patent no 689056 [72]). In particular, PAMAM dendrimers can bind DNA because of their overall positive charge and nano-architecture (US patent no. 0130191). For these reasons, PAMAM dendrimers have been commercialised as plasmid vectors in cell transfection (Superfect™, Quiagen). Recently, dendrons have been used to increase the affinity of specific bioligands to cell receptors by functionalising the last branching of the dendrimer with the targeted bioligand [73]. A patent application has been filed by the University of Brighton in which bi-functional, bio-tethered dendrons are used to functionalise the surface of medical implants (PCT/GB2007/050741) (Fig. 3.5).

This patent application ensures the correct orientation of the dendron at material surfaces, thus maximising the exposure of cell bioligands and docking sites for morphogens and growth factors and increasing their density for area unit considered. These nanostructured biomaterials could therefore be considered and termed as

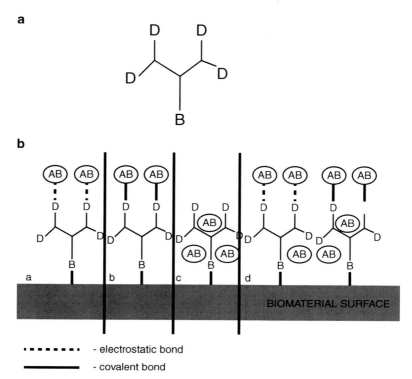

Fig. 3.5 Schematic representation of dendrons for biomaterial surface functionalisation according to patent application PCT/GB2007/050741. (**a**) Single bi-functional G1 dendron, (**b**) biomaterial surface functionalisation with bi-functional dendrimers exposing bioactive molecules with different strategies: from *left* to *right* electrostatic interaction, covalent interactions, entrapment, combination of all. B: Bridging functionality able to recognise and bind the surface of a given biomaterial. D: Docking site for bioactive molecules, (AB): bioactive molecule. Adapted from patent application PCT/GB2007/050741

synthetic pro-morphogens. In their functions these macromolecules would be potentially beneficial both to cell adhesion as they can lead to the spaced distribution of the bioligands interacting with the cell membrane receptors (Fig. 3.1, lower pathway) and to the generation of domains of high concentration of morphogens and growth factors (Fig. 3.1, upper pathway). In the case of cell bioligands, their presentation through an ordered branching of the dendrons is also likely to generate a surface nano-topography able to enhance cell adhesion. Docking sites can be obtained, for example, through the functionalisation of biomaterial scaffold by heparan sulphate oligomers derived from the ECM as those which have been reported to regulate morphogen gradients in developing tissues (see Sect. 3.3) [11]. Indeed, heparin has already been shown to be able to form macromolecular complexes with VEGF and to play a fundamental role in controlling tissue angiogenesis [74].

3.9 Conclusions

Knowledge deriving from the biochemical signalling controlling tissue morphogenesis in embryonic and adult tissues offers the potential for the development of new therapeutic tools in regenerative medicine. To facilitate and control the cell-to-cell communication pathways, peptides and polymers can be synthesised that are capable of exposing to the damaged tissue either specific cell bioligands or functional groups able to bind bioactive molecules. These macromolecules would thus act as biomimetic pro-morphogens contributing to the generation of biochemical signalling gradients (Fig. 3.1, lower and upper pathways). In the first case, biomaterial scaffolds would be tethered with bioligands specifically encouraging the colonisation of "intermediate organisers" able to process morphogens/growth factors secreted by the "organiser" cells (Fig. 3.1, lower pathway). In the second case, the exposure of functional groups of natural or synthetic origin able to capture endogenous morphogens could contribute to the establishment of biochemical gradients (Fig. 3.1, upper pathway). Finally, in the clinical cases where tissue regeneration potential has significantly been comporomise, the use of these so-called synthetic pro-morphogens could be combined with that of specific peptidic growth factor analogues to generate ex novo completely synthetic gradients.

References

1. Kerszberg M, Wolpert L (1998) J Theor Biol 191:103–114
2. Vincent JP, Briscoe J (2001) Morphogens. Curr Biol 11:R851–R854
3. Kornberg TB, Guha A (2007) Understanding morphogen gradients: a problem of dispersion and containment. Curr Opin Genet Dev 17:264–271
4. Wolpert L (2011) Positional information and patterning revisited. J Theor Biol 269:359–365
5. Reddi A (1997) Bone morphogenetic proteins: an unconventional approach to isolation of first mammalian morphogens. Cytokine Growth Factor Rev 8(11):20
6. Pagès F, Kerridge S (2000) Trends Genet 16:40–44
7. Yan D, Lin X (2009) Shaping morphogen gradients by proteoglycans. Cold Spring Harb Perspect Biol 1(3):a002493
8. Ibanes M, Belmonte JCI (2008) Theoretical and experimental approaches to understand morphogen gradients. Mol Syst Biol 2008:176
9. Gregor T, Tank DW, Wieschaus EF, Bialek W (2007) Cell 130:153–164
10. Kerszberg M (2004) Curr Opin Genet Dev 14:440–445
11. Zhu AJ, Scott MP (2004) Genes Dev 18:2985–2997
12. Lander AD, Nie Q, Wan FY (2002) Dev Cell 2:785–796
13. Kerszberg M, Wolpert L (2007) Specifying positional information in the embryo: looking beyond morphogens. Cell 130:205–209
14. Wolpert L, Jessell T, Lawrence P, Meyerowitz E, Robertson E, Smith J (2006) Principles of development, 3rd edn. Oxford University Press, New York, p 24
15. Green JBA, Smith JC (1991) Growth factors as morphogens. Trends Genet 7:245–250
16. Esteve P, Bovolenta P (2006) Secreted inducers in vertebrate eye development: more functions for old morphogens. Curr Opin Neurobiol 16:13–19
17. Zou Y, Lyuksyutova AI (2007) Morphogens as conserved axon guidance cues. Curr Opin Neurobiol 17:22–28

18. Santin M (2009) Towards high-performance and industrially-sustainable tissue engineering products (Chapter 16). In: Santin M (ed) Strategies in regenerative medicine: integrating biology with material science. Springer, New York, pp 467–493, ISBN-10: 0387746595, ISBN-13: 9780387746593

19. Hodde JP, Johnson CE (2007) Extracellular matrix as a strategy for treating chronic wounds. Am J Clin Dermatol 8:61–66

20. Werner S, Krieg T, Smola H (2007) Keratinocyte-fibroblast interactions in wound healing. J Invest Dermatol 127:998–1008

21. Li HH, Fu XB, Zhang L, Huang QJ, Wu ZG, Sun TZ (2008) Research of PDGF-BB gel on the wound healing of diabetic rats and its pharmacodynamics. J Surg Res 145:41–48

22. Lee SH, Shin H (2007) Matrices and scaffolds for delivery of bioactive molecules in bone and cartilage tissue engineering. Adv Drug Deliv Rev 59:339–359

23. Lal S, Barrow RE, Wolf SE, Chinkes DL, Hart DW, Heggers JP et al (2000) Biobrane improves wound healing in burned children without increased risk of infection. Shock 14:314–318

24. Lukish JR, Eichelberger MR, Newman KD et al (2001) The use of bioactive skin substitute decreases length of stay for pediatric burn patients. J Pediatr Surg 36:1118–1121

25. Unezaki S, Yoshii S, Mabuchi T, Saito A, Ito S (2009) Effects of neurotropic factors on nerve regeneration monitored by in vivo imaging in thy1-YFP transgenic mice. J Neoursci Methods 178:308–315

26. Piquilloud G, Christen T, Pfister LA, Gander B, Papaloizos Y (2007) Variations in glial cell line-derived neurotrophic factor release from biodegradable nerve conduits modify the rate of functional motor recovery after rat primary nerve repairs. Eur J Neurosci 26:1109–1117

27. Barras FM, Pasche P, Bouche N, Aebischer P, Zurn AD (2002) Glial cell line-derived neurotrophic factor released by synthetic guidance channels promotes facial nerve regeneration in the rat. J Neurosci Res 70:746–755

28. Seeherman H, Wozney JM (2005) Delivery of bone morphogenetic proteins for orthopedic tissue regeneration. Cytokine Growth Factor Rev 16:329–345

29. Kelpke SS, Zinn KR, Rue LW, Thompson JA (2004) Site-specific delivery of acidic fibroblast growth factor stimulates angiogenic and osteogenic response in vivo. J Biomed Mater Res A 71A:316–325

30. Mabilleau G, Aguado E, Stancu IC, Cincu C, Basle ME, Chappard D (2008) Effects of FGF-2 release from a hydrogel polymer on bone mass and microarchitecture. Biomaterials 29:1593–1600

31. Schmoekel HG, Weber FE, Schense JC, Gratz KW, Schawalder P, Hubbell JA (2005) Bone repair with a form of BMP-2 engineered for incorporation into fibrin cell ingrowth matrices. Biotechnol Bioeng 89(3):253–262

32. Jeon O, Song SJ, Kang SW, Putnam AJ, Kim BS (2007) Enhancement of ectopic bone formation by bone morphogenetic protein-2 released from a heparin-conjugated poly(L-lactic-co-glycolic acid) scaffold. Biomaterials 28:2763–2771

33. Molek P, Strukelj B, Bratkovic T (2011) Peptide phage display as a tool for drug discovery: targeting membrane receptors. Molecules 16:857–887

34. Smith GP (1985) Filamentous fusion phage: novel expression vectors that display cloned antigens on the virion surface. Science 228:1315–1317

35. Smith GP, Petrenko VA (1997) Phage display. Chem Rev 97:391–410

36. Ja WW, Roberts RW (2005) G-protein-directed ligand discovery with peptide combinatorial libraries. Trends Biochem Sci 20:318–324

37. Tipps ME, Lawshe JE, Ellington AD, Mihic SJ (2010) Identification of novel specific allosteric modulators of the glycine receptor using phage display. J Biol Chem 285:22840–22845

38. De Berardinis P, Haigwood N (2004) New recombinant vaccines based on the use of prokaryotic antigen-display systems. Expert Rev Vaccines 3:673

39. Skerra A (2007) Alternative non-antibody scaffolds for molecular recognition. Curr Opin Biotechnol 18:295–304

40. Hertveldt K, Belien T, Volckaert G (2009) General M13 phage display: M13 phage display in indentification and characterization of protein-protein interactions. Methods Mol Biol 502:321–339

41. Sedlacek R, Chen E (2005) Screening for protease substrate by polyvalent phage display. Comb Chem High Throughput Screen 8:197–203
42. Fernandez-Gacio A, Uguen M, Fastrez J (2003) Phage display as a tool for the directed evolution of enzymes. Trends Biotechnol 21:408–414
43. Deller MC, Jones YE (2000) Cell surface receptors. Curr Opin Struct Biol 10:213–219
44. Yamada N, Yanai R, Kawamoto K, Takashi Nagano T, Nakamura M, Inui M, Nishida T (2006) Promotion of corneal epithelial wound healing by a tetrapeptide (SSSR) derived from IGF-1. Invest Ophthalmol Vis Sci 47:3286–3292
45. Brennand DM, Dennehy U, Ellis V, Scully MF, Tripathi P, Kakkar VV, Patel G (1997) Identification of a cyclic peptide inhibitor of platelet-derived growth factor-BB receptor-binding and mitogen-induced DNA synthesis in human fibroblasts. FEBS Lett 413:70–74
46. Hardy B, Raiter A, Weiss C, Kaplan B, Tenenbaum A, Battler A (2007) Angiogenesis induced by novel peptides selected from a phage display library by screening human vascular endothelial cells under different physiological conditions. Peptides 28:691–701
47. D'Andrea LD, Del Gatto A, Pedone C, Benedetti E (2006) Peptide-based molecules in angiogenesis. Chem Biol Drug Des 67:115–126
48. Suzuki Y, Tanihara M, Suzuki K, Saitou A, Sufan W, Nishimura Y (2000) Alginate hydrogel linked with synthetic oligopeptide derived from BMP-2 allows ectopic osteoinduction in vivo. J Biomed Mater Res 50:405–409
49. Horii A, Wang X, Gelain F, Zhang S (2007) Biological designer self-assembling peptide nanofibre scaffolds significantly enhance osteoblast proliferation, differentiation and 3-D migration. PLoS One 2(2):E190. doi:10.1371/journal.pone.0000190
50. Simoes DCM, Vassilakopoulos T, Toumpanakis D, Petrochilou K, Roussos C, Papapetropoulos A (2008) Angiopoietin-1 protects against airway inflammation and hyperreactivity in asthma. Am J Respir Critic Care Med 177:1314–1321
51. Smadja DM, Laurendeau I, Avignon C, Vidaud M, Aiach M, Gaussem P (2006) The angiopoietin pathway is modulated by PAR-1 activation on human endothelial progenitor cells. J Thromb Haemost 4:2051–2058
52. Pollock JF, Healy KE (2009) Biomimetic and bioresponsive materials in regenerative medicine (Chapter 4). In: Santin M (ed) Strategies in regenerative medicine: integrating biology with material science. Springer, New York, pp 97–154, ISBN-10: 0387746595, ISBN-13: 9780387746593
53. Hern DL, Hubbell JA (1998) Incorporation of adhesion peptides into nonadhesive hydrogels useful for tissue resurfacing. J Biomed Mater Res 39:266–276
54. Schense JC, Hubbell JA (2000) Three-dimensional migration of neurites is mediated by adhesion site density and affinity. J Biol Chem 275:6813–6818
55. Shin H, Jo S, Mikos AG (2002) Modulation of marrow stromal osteoblast adhesion on biomimetic oligo[poly(ethylene glycol) fumarate] hydrogels modified with Arg-Gly-Asp peptides and a poly(ethyleneglycol) spacer. J Biomed Mater Res 61:169–179
56. Harbers GM, Healy KE (2005) The effect of ligand type and density on osteoblast adhesion, proliferation, and matrix mineralization. J Biomed Mater Res A 75:855–869
57. Brandley BK, Schnaar RL (1988) Covalent attachment of an Arg-Gly-Asp sequence peptide to derivatizable polyacrylamide surfaces: support of fibroblast adhesion and long-term growth. Anal Biochem 172:270–278
58. Shin H, Zygourakis K, Farach-Carson MC et al (2004) Attachment, proliferation, and migration of marrow stromal osteoblasts cultured on biomimetic hydrogels modified with an osteopontin-derived peptide. Biomaterials 25:895–906
59. Rezania A, Healy KE (1999) Biomimetic peptide surfaces that regulate adhesion, spreading, cytoskeletal organization, and mineralization of the matrix deposited by osteoblast-like cells. Biotechnol Prog 15:19–32
60. Silva GA, Czeisler C, Niece KL et al (2004) Selective differentiation of neural progenitor cells by high-epitope density nanofibers. Science 303:1352–1355
61. Schense JC, Bloch J, Aebischer P et al (2000) Enzymatic incorporation of bioactive peptides into fibrin matrices enhances neurite extension. Nat Biotechnol 18:415–419

62. Nilsson F, Tarli L, Viti F, Neri D (2000) The use of phage display for the development of tumour targeting agents. Adv Drug Deliv Rev 43:165–196
63. Ivanenkov VV, Felici F, Menon AG (1999) Uptake and intracellular fate of phage display vectors in mammalian cells. Biochim Biophys Acta 1448:450–462
64. Stacey A, Palasek ZJ, Cox ZJ, Collins JM (2007) Limiting racemization and aspartimide formation in microwave-enhanced Fmoc solid phase peptide synthesis. J Pept Sci 13:143–148
65. Skerra A (2000) Engineered protein scaffolds for molecular recognition. J Mol Recognit 13:167–187
66. Wurch T, Lowe P, Caussanel V, Bes C, Beck A, Corvaia N (2008) Development of novel protein scaffolds as alternatives to whole antibodies for imaging and therapy: status on discovery research and clinical validation. Curr Pharm Biotechnol 9:502–509
67. Duncan R, Izzo L (2005) Dendrimer biocompatibility and doxicity. Adv Drug Deliv Rev 57:2215–2237
68. Esfand T (2001) Poly(amidoamine) (PAMAM) dendrimers: from biomimicry to drug delivery and biomedical applications. Drug Discov Today 6:427–436
69. Hobson LJ, Feast WJ (1999) Poly(amidoamine) hyperbranched systems: synthesis, structure and characterization. Polymer 40:1279–1297
70. Wells NJ, Basso A, Bradley M (1999) Solid-phase dendrimer synthesis. Biopolymers 47:381–396
71. Al Jamal KT, Ramaswamy C, Florence AT (2005) Supramolecular structures from dendrons and dendrimers. Adv Drug Deliv Rev 57:2238–2270
72. Tang MX, Redemann CT, Szoka FC (1996) In vitro gene delivery by degraded polyamidoamine dendrimers. Bioconjug Chem 7:703–714
73. Monaghan S (2001) Solid-phase synthesis of peptide-dendrimer conjugates for an investigation of integrin binding. ARKIVOC 46–53
74. Wijelath E, Namekata M, Murray J, Furuyashiki M, Zhang S, Coan D, Wakao M, Harris RB, Suda Y, Wang LM (2010) Multiple mechanisms for exogenous heparin modulation of vascular endothelial growth factor activity. J Cell Biochem 111:461–468

Chapter 4
The Role of Oxidative Stress in the Response of Endothelial Cells to Metals

Roman Tsaryk, Kirsten Peters, Ronald E. Unger, Dieter Scharnweber, and C. James Kirkpatrick

4.1 Metallic Materials for Implantation

Metallic biomaterials have been used for many decades in the production of various devices for medical application. Due to their excellent mechanical properties metallic materials are used in orthopaedic prostheses for bone and joint replacement. They are, for instance, utilised in total hip and knee arthroplasty, spine surgery, in craniofacial and maxillofacial reconstruction and as the load-bearing parts of bone implants, such as pins and plates, for the fixation of fractured bones. Apart from bone- and joint-related applications dental implants and cardiovascular devices (pacemakers, intravascular stents, etc.) are produced from metals and their alloys.

The most common metallic materials used in implantology are stainless steel, CoCr-based alloys, commercially pure Ti (cpTi) and titanium alloys, such as Ti6Al4V, Ti6Al7Nb, Ti5Al2.5Fe and others. In joint replacement metallic materials are used either alone or in combination with other materials, such as ceramics, ultra-high-molecular-weight polyethylene, polyurethane, cement, etc. Different metals and their alloys are used for different applications based on their properties. While stainless steel and CoCr-based alloys are known for their mechanical strength and stiffness as well as high wear resistance [56] Ti-based alloys usually display higher

R. Tsaryk (✉) • R.E. Unger • C. J. Kirkpatrick
Institute of Pathology, University Medical Center of the Johannes Gutenberg University,
Langenbeckstr. 1, Mainz 55131, Germany
e-mail: tsaryk@uni-mainz.de; runger@uni-mainz.de; kirkpatrick@ukmainz.de

K. Peters
Department of Cell Biology, University of Rostock, Schillingallee 69, Rostock 18057, Germany
e-mail: kirsten.peters@med.uni-rostock.de

D. Scharnweber
Max Bergman Center of Biomaterials, Technische Universität Dresden,
Budapester Str. 27, Dresden 01069, Germany
e-mail: Dieter.Scharnweber@tu-dresden.de

I. Antoniac (ed.), *Biologically Responsive Biomaterials for Tissue Engineering*,
Springer Series in Biomaterials Science and Engineering 1,
DOI 10.1007/978-1-4614-4328-5_4, © Springer Science+Business Media New York 2013

biocompatibility [69]. Other advantages of Ti-containing alloys include their relative lightness (high strength-to-weight ratio), higher flexibility, fatigue resistance, capacity for joining with bone (osseointegration) and corrosion resistance [18, 50]. The latter is attributed to the fact that the surface of titanium (-alloy) implant is covered with a titanium oxide (TiO_2) film which drastically reduces the corrosion rate of the metal [94]. It is known, however, that Ti-based materials exert a high mechanical friction effect that might lead to rupture or weakening of the TiO_2 layer, thus inducing corrosion processes and the formation of wear debris in such regions. Therefore, the good mechanical properties of metals and their alloys regarding manufacture and stability of metal implants are often overshadowed by the cases of adverse biological reactions that might cause implant failure and the requirement for revision surgery.

4.2   Aseptic Loosening and Metal Corrosion

Aseptic loosening is one of the main reasons for the failure of orthopaedic implants. In fact, over 25 % of all orthopaedic implants show features of aseptic loosening [96]. It is characterised by the bone loss (osteolysis) around the prosthesis and appearance of a synovial-like peri-prosthetic membrane reminiscent of a foreign body reaction, resulting in implant loosening that often requires revision surgery [31]. The failure of metal implants is attributed mainly to the formation of wear debris and corrosion of the implant materials, though cyclic loading, stress shielding and hydrodynamic pressure around the implant also play their role [1, 19, 33, 34, 41, 53, 82]. As soon as load is applied to metal devices, they are permanently exposed to the action of mechanical forces and chemical attack. Micromotions and the friction between components of the implant as well as the implant and surrounding tissues are believed to be the main causes of the particulate wear debris generation [27, 33, 82]. The effects of the metal particles can be augmented by polyethylene particles in the case of metal-on-polyethylene implants and cement wear debris, originating from materials used for implant fixation [1, 82]. Particles can often be detected in the peri-implant tissues as well as in remote locations. CoCr particles can be found in local and distant lymph nodes, bone marrow, liver and spleen of subjects with hip implants with higher particle amounts detected in subjects with loosened implants [7].

Metal degradation products can induce an inflammatory response in the peri-implant region [33]. This can occur during the initial phase of implant integration but also happens over the lifetime of the implant if wear and corrosion products are continuously developing. Such chronic inflammation may lead to metal surface modifications due to the interactions with soluble mediators, mainly ROS, extensive formation of granulation tissue and fibrous capsule, peri-prosthetic bone resorption as a result of osteoclast activation by inflammatory stimuli and finally aseptic loosening of the implant. An important role in bone resorption is played by the TNF-related cytokine, receptor activator of nuclear factor kappa B ligand (RANKL).

Interaction of RANK-positive monocytes with RANKL-positive fibroblasts or osteoblasts is known to lead to multi-nuclear giant-cell formation and osteoclast differentiation from progenitor macrophages [41]. RANKL interaction with its receptor is inhibited by osteoprotegerin (OPG). Endothelial cells are the main producers of OPG, although in the case of synovial membrane formation they are located remotely from bone resorption sites, further pointing to the important role of endothelial cells in aseptic loosening [55]. TNF-α and IL-1β are believed to regulate the RANKL system, although the direct influence of metal degradation products on osteoclastogenesis has also been reported. Thus, macrophages that phagocytosed metal and polymer particles were shown to differentiate into osteoclasts capable of bone resorption [73]. Furthermore, metals undergo electrochemical corrosion, which can lead to release of metal ions [33]. Metal particles are also a source of metal ions, therefore amplifying tissue responses to metal degradation products. This response can include local and remote toxicity, induction of chronic inflammation, bone resorption and loosening and failure of implants [1, 53, 82]. Although considerable information has been collected since the onset of metal implant use, the cellular reactions induced by implant degradation products that may finally lead to implant failure are not yet well studied.

Metal corrosion is an electrochemical process occurring at the interface between metal and its environment. It consists of anodic and cathodic half-reactions [101]. The two half-reactions can be separated spatially. Metal oxidation takes place in the anodic region, and the electrons released in this process flow to the cathodic region, where they take part in the oxygen reduction in aqueous solutions with neutral pH. In acidic pH cathodic half-reaction leads to hydrogen development. While anodic half-reaction usually takes place at the sites of metal structure irregularities or mechanical damage, the cathodic half-reaction can occur almost anywhere due to the conductor properties of metals and the electrolyte-containing body liquids.

As a result of anodic half-reaction, metal ions are released into solution (4.1). This process also occurs in vivo and is manifested as increased concentration of metal ions in the serum of patients with metal implants [29]. Furthermore, an oxide layer is formed on the surface of the metal as a result of this process (4.2).

Anodic Reaction

$$\text{Metal dissolution}: \ \text{Me} \rightarrow \text{Me}^{z+} + z e^-, \qquad (4.1)$$

$$\text{Oxide formation}: \ \text{Me}^{z+} + \frac{3}{2} z \text{H}_2\text{O} \rightarrow \text{MeO}_{z/2} + z \text{H}_3\text{O}^+. \qquad (4.2)$$

In contrast, cathodic partial reaction and the effects of its products on biological processes are rarely taken into consideration in experimental studies. Importantly, however, cathodic oxygen reduction (4.3 and 4.4) proceeds through several partial reactions [4], resulting in oxygen radicals and hydrogen peroxide as intermediate products as indicated in (4.5–4.8).

Cathodic Reaction

Hydrogen development (pH < 4):

$$2H_3O^+ + 2e^- \rightarrow H_2 \uparrow + 2H_2O. \tag{4.3}$$

Oxygen reduction (pH > 4):

$$O_2 + 2H_2O + 4e^- \rightarrow 4OH^-, \tag{4.4}$$

$$O_2 + H_2O + e^- \rightarrow HO_2^{\bullet} + OH^-, \tag{4.5}$$

$$HO_2^{\bullet} + H_2O \rightarrow H_2O_2 + {}^{\bullet}OH, \tag{4.5a}$$

$${}^{\bullet}OH + e^- \rightarrow OH^-, \tag{4.6}$$

$$H_2O_2 + e^- \rightarrow {}^{\bullet}OH + OH^-, \tag{4.7}$$

$${}^{\bullet}OH + e^- \rightarrow OH^-. \tag{4.8}$$

Formation of ROS and H_2O_2 on the metal implant surface during cathodic half-reaction of corrosion may have detrimental effects on surrounding tissues and contribute to the complex processes eventually leading to aseptic loosening.

Formation of metal oxide at the interface with the environment is another important result of anodic half-reaction. This usually terminates the metal corrosion process. Oxide formation is especially important for Ti and Ti-based alloys. The titanium oxide layer (mainly consisting of TiO_2) is known to be relatively stable, and this is believed to be the reason for inactivity of Ti-based materials and their well-documented biocompatibility. Experimental thickening of TiO_2 layer was shown to lead to a decreased release of Ti ions [46]. However, disruption of the TiO_2 layer immediately renews the corrosion process, resulting in further metal ion release and ROS formation. Such TiO_2 layer disruption happens when the implant is exposed to friction with other implant parts or surrounding tissues, especially with bone tissue. Metal debris induces abrasive wear, causing further TiO_2 layer breakdown on the implant surface, thus amplifying the corrosion process [50]. Although TiO_2 is considered to be relatively inactive, ROS and H_2O_2 formed in high quantities by activated inflammatory cells can modify the TiO_2 layer, since H_2O_2 in a phosphate-buffered solution leads to pronounced thickening of the TiO_2 layer [60].

Activated macrophages that produced high levels of ROS accelerated Ti ion release from cpTi, pointing to ROS-induced corrosion [62]. All this indicates the possibility of metal corrosion occurring even on materials considered to be relatively inactive from a biological point of view.

4.3   Integration of Metal Implants and the Role of Endothelial Cells

Upon implantation tissues around the implant respond with reactions similar to the wound healing process. This process can be divided into three phases—inflammatory, proliferating and remodelling or maturation phase [28, 48, 57, 60]. During the first phase inflammatory cells, such as neutrophils and monocytes, are recruited to the site of implantation to remove cellular debris and possible sources of infection [93]. The inflammatory phase is followed by the proliferative phase, characterised by the formation of granulation tissue. It consists mainly of fibroblasts that proliferate and produce extracellular matrix (ECM). To supply the newly formed tissue with nutrients and oxygen, vascularisation is induced in the form of granulation tissue [17]. Tissue vascularisation at the interface of an orthopaedic implant is crucially important for implant integration and stability. Endothelial cells play a central role in the formation of blood vessels in the newly formed peri-implant tissue in a process called angiogenesis. Moreover, they take part in the inflammatory phase of wound healing after surgery. They do so by releasing cytokines (for example, IL-8 and MCP-1), which are involved in the recruitment of leukocytes to the wound area. Furthermore, adhesion molecules (e.g. E-selectin and ICAM-1) expressed on the surface of endothelial cells allow the attachment and extravasation of inflammatory cells into the peri-implant tissue [47]. Wound healing around an implant is concluded with the maturation or remodelling phase, during which provisional ECM is replaced by mature ECM and the cell density in the peri-implant tissue is reduced [60]. Close proximity of endothelial cells to the surface of metal implants enables their exposure to metal degradation and corrosion products. While the reactions of endothelial cells to metal particles and metal ions have been relatively well studied, their response to electrochemical processes, especially to processes associated with cathodic partial reactions on the metal surface, has not yet been examined.

4.4   Oxidative Stress: ROS Formation and Antioxidant Defence Mechanisms

Molecular oxygen possesses two unpaired electrons with the same spin in the external π^* orbital that is responsible for the ease with which O_2 forms radicals, or ROS [23]. ROS, such as singlet oxygen ($^1O_2$), hydroxyl radical ($^.OH$) and superoxide radical ($^.O_2-$), are a group of free radicals with high reactivity [10]. H_2O_2 is another non-radical product of incomplete oxygen reduction. ROS and H_2O_2 are products of normal cell metabolism and are important for cell signalling but can also be toxic to cells by causing damage to nearly all macromolecules [15]. Cells possess enzymatic and non-enzymatic antioxidants involved in ROS elimination [20]. However, when production of ROS is increased upon certain stimuli, the balance of ROS formation and degradation may be disturbed. Such a state is called oxidative stress.

ROS can be formed in the cell as by-products of electron transfer by a chain of cytochromes in mitochondrial respiration; cytochromes of the P450 family, involved in detoxification of various compounds, also produce ROS. Cellular oxidases, such as NADPH oxidase, myeloperoxidase, glucose oxidase, cyclooxygenase and xanthine oxidase, are another potent ROS source [95]. Oxidative stress can result in damage to most biological molecules. Lipid peroxidation is one of the best known markers of oxidative stress [76]. It comprises a set of chain reactions in which lipid peroxyl radicals are formed from unsaturated fatty acids and induce similar reactions in other fatty acid molecules. Peroxyl radicals can be rearranged via a cyclisation reaction to endoperoxides, which decompose to aldehydes, such as malondialdehyde (MDA) and 4-hydroxy-2-nonenal (HNE) [13, 91]. Other lipids, such as cholesterol, can also undergo oxidation reactions [63]. ROS can also oxidise proteins. The changes in amino acid residue side chains caused by ROS can induce conformational changes in protein structure that eventually can lead to loss of activity and protein misfolding [91]. Cells possess the mechanism to cope with unfolded and misfolded proteins called the unfolded protein response (UPR) [75]. UPR inhibits general translation, but induces gene expression and protein translation of chaperones, such as heat shock proteins (HSP) [64]. Although there are reports connecting oxidative stress to UPR [54] the link between those two processes has been poorly studied.

Cells possess mechanisms to defend themselves against ROS. Natural compounds present in the cell, such as vitamin E, vitamin C, carotenoids, polyphenols, bioflavonoids, selenium, copper, zinc and manganese, can function as non-enzymatic antioxidants [20]. On the other hand, antioxidant enzymes catalyse highly specific reactions which detoxify ROS and organic peroxides (Fig. 4.1). These include antioxidant enzymes, such as superoxide dismutase (SOD), catalase and glutathione peroxidise (GPX), which determine the response of cells to oxidative stress [6]. Two isoforms of SOD, cytosolic and nuclear (in endothelial cells) Cu/Zn-SOD (SOD1) and mitochondrial Mn-SOD (SOD2), catalyse the dismutation of superoxide radicals into H_2O_2 and O_2 by successive oxidation and reduction of the transition metal ion in the active site (4.9) [30]. Interestingly, O_2- can interact with nitric oxide (NO), forming another highly reactive compound peroxinitrite (ONOO$^-$) [91]. H_2O_2 can be degraded by both catalase and GPX. Catalase is a heme-containing enzyme that catalyses hydrolysis of H_2O_2 (4.10), which is localised in peroxisomes and protects cells from H_2O_2 formed during β-oxidation of long-chain fatty acids [13]. GPX, which is mostly a cytosolic enzyme, can catalytically reduce H_2O_2 (4.11) as well as lipid peroxides (4.12) [92]. GPX requires glutathione (γ-glutamyl-cysteinyl-glycine, GSH) for these reactions. GSH, in turn, can react with H_2O_2 directly or through a GPX-catalysed reaction. As a result, two GSH molecules are oxidised to form GSSH, which is

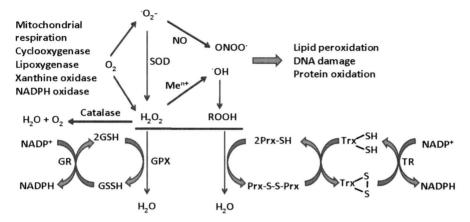

Fig. 4.1 ROS formation and degradation pathways

then reduced by NADPH-dependent glutathione reductase (GR) [97]. GSH is also involved in the reactions catalysed by glutathione-S-transferases (GST), which conjugate GSH to various cellular metabolites and xenobiotics that are eventually converted to mercapturic acids for excretion [38]:

$$2^\bullet O_2^- + 2H^+ \to H_2O_2 + O_2, \tag{4.9}$$

$$2H_2O_2 \to 2H_2O + O_2, \tag{4.10}$$

$$H_2O_2 + 2GSH \to 2H_2O + GSSG, \tag{4.11}$$

$$ROOH + 2GSH \to ROH + H_2O + GSSH. \tag{4.12}$$

In the presence of heavy metal ions H_2O_2 can be converted into highly toxic $^\bullet OH$ via the Fenton (4.9) and Haber–Weiss reactions (4.13–4.15):

$$Me^{n+} + H_2O_2 \to Me^{(n+1)+} + {^\bullet OH} + OH^-, \tag{4.13}$$

$$^\bullet O_2^- + H_2O_2 \to O_2 + {^\bullet OH} + OH^-, \tag{4.14}$$

$$Me^{(n+1)+} + {^\bullet O_2^-} \to Me^{n+} + O_2. \tag{4.15}$$

Peroxiredoxins (Prx) form another family of proteins (consisting of six members in mammalian cells) capable of reducing H_2O_2 and alkylhydroperoxides with the use of thioredoxin (Trx). Prx exist as homodimers with one cysteine residue per active centre. Upon interaction with H_2O_2 cysteine sulfenic acid (Cys-SOH) is formed in the active site of one subunit, which subsequently forms a disulphide bond with an additional cysteine residue of another subunit. The disulphide bond is then reduced by Trx, a protein that has two cysteine groups that form disulphide bonds upon oxidation. Trx, in turn, is reduced by the NADPH-dependent enzyme thioredoxin reductase [22, 32, 71].

Physiological concentrations of ROS are involved in cell signalling. Among others, insulin, VEGF, PDGF and TNF-α-induced signalling is ROS dependent [21, 71]. Protein tyrosine phosphatases (PTP) have been shown to be one of the specific targets of ROS in cell signalling. Their activity is reduced upon ROS oxidation of cysteine residues essential for PTP catalysis and can be restored by cellular thiols. PTP activity regulates, in turn, the signalling cascades induced by receptor tyrosine kinases [22, 45].

ROS play an important role in endothelial cell functions. Vascular NADPH oxidase and endothelial NO-synthase (eNOS) are the important sources of ROS in endothelial cells. ROS can regulate among others inflammatory and angiogenic responses of endothelial cells, thus contributing to wound healing [52, 90]. Oxidative stress has also been shown to be involved in the pathogenesis of cardiovascular diseases, such as atherosclerosis, hypertension and ischemia–reperfusion injury [15].

4.5   Ti6Al4V-Induced Oxidative Stress

Titanium and titanium alloys are materials of choice for many medical applications. Apart from their excellent mechanical properties, widely accepted titanium corrosion resistance and good biocompatibility are the main reasons for wide use of titanium-based biomaterials. The surface of Ti6Al4V, the titanium alloy used in this study, as well as other titanium-based materials, is normally covered with a 3–5 nm thick air-formed TiO_2 layer that is thought to be responsible for their biocompatibility [94]. However, several facts point to the reactivity of titanium surfaces. Bikondoa et al. [2] showed that defects such as oxygen vacancies in a model oxide surface, rutile TiO_2 (110), mediate the dissociation of water. Another indication for titanium alloy reactivity is an elevated concentration of titanium in the serum of patients after implantation. Dramatic increase in titanium serum concentration was demonstrated in patients with failed knee implants [29, 35]. Titanium release is a result of the anodic corrosion process, which takes place at the sites of defects in the TiO_2 layer. Wear particles formed at the interface bone/implant or implant/cement (used for the fixation of the implant) can disrupt the protective TiO_2 film, thus further inducing corrosion and leading to the synergistic fretting-wear process. Metal ions are released from the metal implants as well as from their wear debris and can reach significant concentrations in peri-implant tissues and blood. Ti-ion concentrations in tissues surrounding hip prosthesis can reach 0.6 mM [3]. It is important to note that normally only Ti ion elevation is detected in patients with implants made of titanium alloys. Hence, the concentration of Al showed no differences between

Fig. 4.2 Quantification of the proliferation status of HDMEC grown on PS and Ti6Al4V 24 h after treatment with H_2O_2 as assessed by measuring Ki67 expression related to the cell number (means ± SDs; untreated control on PS set as 100 %, significant difference:*$p<0.05$, **$p<0.01$)

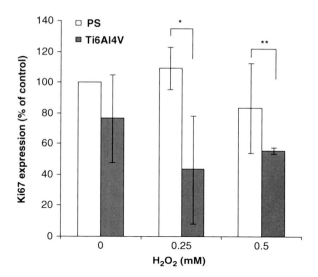

patients with failed and stable Ti6Al4V alloys, while the concentration of V remained low [36]. An increased metal ion concentration can be detected even in individuals with no complications after implantation of Ti-based prostheses, suggesting that metal corrosion occurs even in the absence of metal wear debris formation [37].

The cathodic part of the corrosion process, in contrast, results in the reduction of oxygen at physiological pH with the formation of ROS and H_2O_2 as intermediate products [59]. Thickening of the TiO_2 layer in biological solutions substantiates the fact that corrosion processes permanently occur at titanium implant surfaces [49, 66]. Besides the possible formation of ROS by the titanium (-alloy) itself as the result of cathodic corrosion, titanium may be subjected to the ROS produced by inflammatory cells coming into contact with titanium (-alloy) surfaces directly after implantation. One of the mediators is H_2O_2 released by monocytes, macrophages and granulocytes. The TiO_2 layer may interact with H_2O_2 leading to formation of hydroxyl radicals [44]. Altogether, these facts led to the hypothesis that endothelial cells that take part in wound healing early after implantation might be permanently subjected to ROS formed at the titanium implant surface, which exceeds physiological protection mechanisms and can thus be referred to as oxidative stress. To confirm this hypothesis the reactions of endothelial cells to the Ti6Al4V alloy in the presence of the oxidative stress inducer H_2O_2 in comparison to the reactions elicited in endothelial cells grown on cell culture PS were studied.

The quantification of human dermal microvascular endothelial cells (HDMEC) growing on Ti6Al4V and PS revealed the dose-dependent reduction of cell number to the same extent on PS and Ti6Al4V alloy 24 h after H_2O_2 treatment compared to the untreated control, with no significant differences between the materials [87]. In contrast, expression of Ki67, a marker for cell proliferation, was lower in HDMEC grown on Ti6Al4V alloy compared to the cells grown on PS (Fig. 4.2). This pointed to the possibility of higher H_2O_2 cytotoxicity on Ti6Al4V alloy. It has been shown in

Fig. 4.3 Dose response of HDMEC grown on PS and Ti6Al4V to H$_2$O$_2$ treatment. MTS conversion was measured 24 h after H$_2$O$_2$ addition (means ± SDs; untreated control on PS set as 100 %)

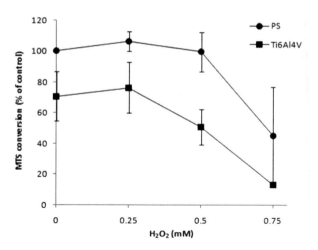

Fig. 4.3 Dose response of HDMEC grown on PS and Ti6Al4V to H$_2$O$_2$ treatment. MTS conversion was measured 24 h after H$_2$O$_2$ addition (means ± SDs; untreated control on PS set as 100 %)

fibroblasts that H$_2$O$_2$ concentrations between 0.12 and 0.4 mM led to growth arrest, whereas higher H$_2$O$_2$ concentrations induced cell death via apoptosis (0.5–1.0 mM) or necrosis (5.0–10.0 mM) (reviewed in [12]). Low H$_2$O$_2$ concentrations (<0.05 mM) were also shown to inhibit proliferation in endothelial cells [14]. The discrepancy in cell number and proliferation ability of HDMEC on Ti6Al4V could possibly be explained by the higher initial attachment of endothelial cells to Ti6Al4V alloy. In fact, it was demonstrated that osteoblasts attach to Ti6Al4V alloy in higher numbers compared to PS, which correlated with a better cell spreading [79]. Another study, however, showed opposite results for epithelial cells [77]. The initially higher number of HDMEC on Ti6Al4V could gradually be reduced as a result of slower proliferation and possible higher cytotoxicity leading to the equal cell numbers on Ti6Al4V and PS at the measurement time point.

The LDH-release cytotoxicity assay revealed a concentration-dependent increase of LDH release from HDMEC after H$_2$O$_2$ treatment [87]. The LDH release indicates damage of the cell membrane, a characteristic feature of necrotic cell death. Moreover, the LDH release was higher in cells grown on Ti6Al4V alloy compared to those in contact with PS. Another indication for a higher degree of cytotoxicity, the reduction of cellular metabolic activity (determined by the MTS viability assay which mirrors the energy metabolic state, e.g. the development of NADH), was also observed after H$_2$O$_2$ treatment. While the H$_2$O$_2$ concentrations tested did not induce major changes in the metabolic activity of HDMEC grown on PS, they significantly decreased MTS conversion in the cells grown on the Ti6Al4V alloy (Fig. 4.3). Importantly, even the basic MTS conversion rates (i.e. in cells not treated with H$_2$O$_2$) were lower in the cells grown on Ti6Al4V alloy compared to the cells grown on PS. It has been shown that oxidative stress induced by H$_2$O$_2$ treatment can lead to a rapid depletion of NADH and ATP due to an increased catabolic rate [74]. Thus, the reduction of metabolic activity in cells grown on Ti6Al4V could be a result of H$_2$O$_2$-induced oxidative stress. However, this does not explain the absence of metabolic reduction of cells in contact with PS. It is possible that compensatory mechanisms

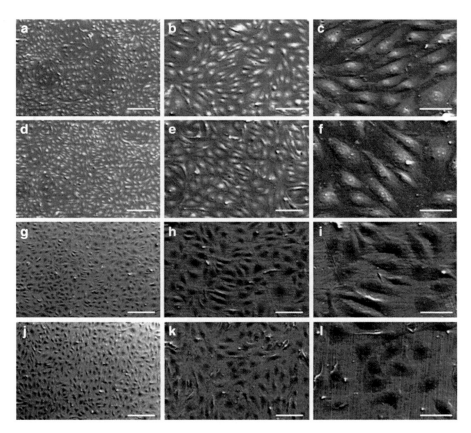

Fig. 4.4 Fig. 4.4 The phenotype of HDMEC grown on PS and Ti6Al4V alloy. HDMEC were cultured on PS (**A-F**) and Ti6Al4V (**G-L**) alloy for 72 h, untreated (**A-C, G-I**) or treated (**D-F, J-L**) with 0.5 mM H2O2 for 24h (H2O2 was added 48 h after seeding of the cells on the materials). Eventually the cells were fixed and scanning electron microscopical images were taken (scale bar: A, D, G, J – 200 µm; B, E, H, K – 100 µm; C, F, I, L – 50 µm).

in HDMEC on PS prevent changes in metabolic activity after H_2O_2 treatment, while Ti6Al4V itself and H_2O_2 induce cumulative effects which result in reduced adaptive potential of the cells.

The phenotype of endothelial cells on Ti6Al4V alloy was visualised with SEM 72 h after seeding the cells on the material and compared to the phenotype of endothelial cells grown on PS (Fig. 4.4). While optically the amount of cells seemed to be similar on both materials, higher magnification images (200× and 500×) indicated the differences in the shape and distribution of HDMEC on PS and Ti6Al4V. Thus, HDMEC grown on PS had an almost uniform spindle-like form lying roughly parallel and forming an almost confluent cell layer. HDMEC on Ti6Al4V alloy, in contrast, had diverse cell shapes with some cells being more rounded and some more elongated than the cells on PS. Cell monolayers on Ti6Al4V alloy had gaps between the cells and lacked the ordered alignment of HDMEC on PS with the cells growing in different directions.

To examine the effects of H_2O_2 on the phenotype of endothelial cells H_2O_2 was added for 24 h to HDMEC growing on PS and Ti6Al4V alloy. 0.5 mM H_2O_2 did not induce marked changes in the phenotype of HDMEC on PS (Fig. 4.4). The cells retained their spindle-like shape, organised alignment and the degree of confluence. HDMEC grown on Ti6Al4V alloy appeared to be more influenced by H_2O_2. The alignment of the cells became more disordered, most of the cells lost the spindle-like shape and appeared rounded, and the amount and the size of the gaps in the cell monolayer increased. In general this points to the higher toxicity of H_2O_2 to HDMEC grown on Ti6Al4V alloy, compared to HDMEC on PS.

As already mentioned, H_2O_2 can induce oxidative stress in endothelial cells by promoting ROS formation. This is achieved by several mechanisms, including activation of ROS production by mitochondria, NADPH oxidases, xanthine oxidase, uncoupled eNOS, etc. [6]. H_2O_2 in the cell can also undergo the Fenton reaction in the presence of metal ions, resulting in the formation of highly toxic hydroxyl radicals. Using the DCF-assay higher levels of ROS was observed in endothelial cells grown on Ti6Al4V alloy 1 h after H_2O_2 addition compared to the cells on PS [87]. DCF is the resulting oxidation product from DCDHF, which is oxidised by peroxyl radical, peroxynitrite and also H_2O_2 [25]. Therefore, increases in DCF-fluorescence can be explained through oxidation by H_2O_2, which enters the cell. However, since the H_2O_2 concentration remaining in the medium 1 h after its addition is nearly similar on both materials the higher DCF-fluorescence in cells grown on Ti6Al4V alloy might reflect further ROS production. One possible explanation for this could be the formation of ROS due to the reactivity of H_2O_2 with the TiO_2 layer. A study by Lee et al. [44] showed that ·OH were formed during the interaction of TiO_2 and H_2O_2 in vitro, while UV-irradiated TiO_2 reacting with H_2O_2 could produce ·O_2^-. Other studies, on the contrary, demonstrate no formation of ·OH in the reaction between H_2O_2 and TiO_2 [84, 85]. Tengvall et al. suggested that a Ti-peroxy gel is formed as a result of the interaction of Ti with H_2O_2 [85]. This gel is believed to trap ·O_2^- [84]. Interestingly, in the absence of serum, Ti-peroxy gels could decrease ROS production by activated leukocytes [43]. Opposite effects of Ti-peroxy gel were also reported. Thus, while the TiO_2 layer was acting as a scavenger for ·OH, the Ti-peroxy gel could amplify detrimental effects of ·OH on ECM proteins, as well as cause these effects alone in vitro [83]. High amounts of H_2O_2 are needed to produce Ti-peroxy gel in vitro. Such conditions are probably unachievable in vivo, although the formation of Ti-peroxy gel and trapping of radicals could hypothetically contribute to the tissue reactions to the implant over longer time periods. This may also be a reason for a relatively better performance of Ti materials compared to other metal materials. However, higher ROS formation in HDMEC on Ti6Al4V alloy shown in this study favours the interaction of H_2O_2 with the TiO_2 layer on the alloy surface. It is also known that traces of iron are present in titanium alloys (according to ASTM F136 Ti6Al4V is allowed to have up to 0.2 % Fe). The presence of iron in titanium

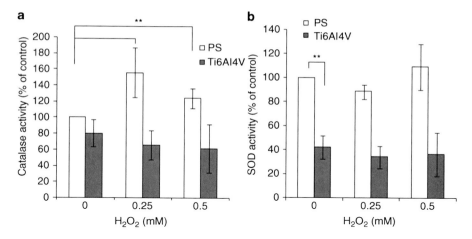

Fig. 4.5 Measurement of antioxidant enzyme activity in HDMEC grown on PS and Ti6Al4V. Catalase (**a**) and SOD (**b**) activities were measured in HDMEC 4 h after H_2O_2 treatment with corresponding activity assay and related to protein concentration of the samples. The data are presented as % of control (untreated HDMEC on PS; means ± SDs; significant difference: **$p < 0.01$)

alloy surface could also lead to the formation of ROS due to the Fenton reaction upon H_2O_2 exposure.

Cells have several mechanisms to protect themselves against oxidative stress. SOD, catalase and GSH system are the central defence players in enzymatic ROS detoxification. The balance between GSH and its oxidised form (GSSG) is critical for protecting cells against oxidants. Excessive formation of GSSG and glutathione conjugates with organic molecules during prolonged oxidative stress leads to the depletion of the GSH pool [16]. Thus, the reduced GSH concentrations in endothelial cells grown on Ti6Al4V alloy compared to the cells grown on PS [87] could reflect a state of permanent oxidative stress in cells in contact with Ti6Al4V alloy as a result of ROS formation on the titanium surface. This is supported by the fact that the GSH concentration in endothelial cells on Ti6Al4V was reduced by H_2O_2 treatment, whereas in cells grown on PS the H_2O_2 treatment induced an elevation of GSH level. The effects of H_2O_2 treatment on PS are in agreement with reports of adaptive increase in GSH concentration after exposure to H_2O_2, a fact that can be explained by the onset of GSH de novo synthesis and H_2O_2-induced up-regulation of the expression of GSH synthesis enzymes (γ-glutamylcysteine synthetase and GSH synthetase) by these temporary oxidative stimuli [51, 70].

Additional evidence for permanent oxidative stress on Ti6Al4V is based on the analysis of enzymatic activities of catalase and SOD. Cells grown on Ti6Al4V alloy showed significantly lower SOD activities than cells grown on PS (Fig. 4.5b). This could be explained by a persistent ROS formation in cells on the Ti6Al4V surface resulting in the exhaustion of enzyme activity, e.g. due to protein oxidation by ROS [11]. Catalase activity in endothelial cells grown on Ti6Al4V was only slightly reduced compared to cells cultured on PS (Fig. 4.5a). Moreover, H_2O_2 treatment of endothelial cells grown on PS elevated catalase and SOD activity. It is known that oxidative stress may increase the activity of antioxidant enzymes. The mechanism of this induction may include changes in enzymatic activity of existing proteins and de novo enzyme synthesis [58]. H_2O_2 exposure induced a slight decrease in the activity of both enzymes in endothelial cells grown on Ti6Al4V. This reduction in SOD and catalase activity could be caused by a cumulative effect of the temporary H_2O_2-induced and the permanent Ti6Al4V-induced oxidative stress. Consequently, the facts demonstrated above indicated an elevated sensitivity of endothelial cells to oxidative stress when growing in contact with Ti6Al4V and thus a reduced antioxidant defence potential.

An animal study by Ozmen et al. [65] suggested a similar effect in vivo: it was shown that 1 month after the insertion of titanium implants in rabbits an increase in lipid peroxidation (a possible result of oxidative stress) in the tissues surrounding the titanium implants occurred, whereas there was a decrease in the activities of antioxidant enzymes (e.g. catalase, SOD, GPX). The same reactions, though to a lesser extent, were seen in the tissues around implanted stainless steel but were minimal in the case of implanted polyethylene. These observations indicate a state of permanent oxidative stress at the surface of titanium implants, leading to exhaustion of antioxidant enzymes in the surrounding tissues. In another study, increased oxidative stress was observed in patients in the tissue surrounding stable and loose total hip implants [40]. It was manifested by a low GSH/GSSH ratio and elevated levels of MDA, a known product of lipid peroxidation, in patients compared to control individuals. Increased H_2O_2 concentrations and decreased catalase activity were also detected in the fibrotic capsule from a failed joint implant [89]. Endothelial cells were also shown to undergo oxidative stress when growing in contact with a NiTi alloy. However, this was attributed to the presence of the transition metal Ni in the TiO_2 layer [68]. Vanadium, which is a component of the alloy used in this study, is known to be toxic and to induce formation of ROS via Fenton and Haber–Weiss reactions [99]. However, on analysis of the surface oxide film of the Ti6Al4V alloy using X-ray photoelectron spectroscopy vanadium oxide was not detected [61], but the presence of traces of vanadium on the surface of titanium alloy cannot be completely excluded.

Another important function of endothelial cells is their role as the driving cell type in angiogenesis. The influence of titanium degradation products on angiogenic potential of endothelial cells has not yet been examined. In this study there were no differences seen in angiogenic potential of HDMEC in an in vitro assay performed on PS or Ti6Al4V (Fig. 4.6). It has to be stated, however, that the cells were not growing in direct contact to the metal surface. Thus, the effects of oxidative stress on Ti6Al4V could be masked by collagen and fibrin used in the assay. Addition of 0.5 mM H_2O_2, however, reduced the formation of tube-like structures on both materials. There are contradictory reports in the literature regarding the effects of H_2O_2 on endothelial cells. In one study the low doses of H_2O_2 (0.1 or 1 µM) induced proliferative activity of endothelial cells and increased tube formation in an in vitro angiogenesis assay. A higher concentration of H_2O_2 (10 µM), in turn, reduced the

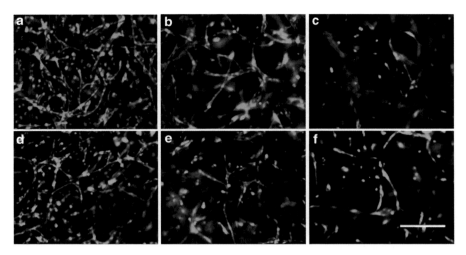

Fig. 4.6 Formation of in vitro capillaries by HDMEC embedded in fibrin/collagen type I gel polymerised on top of PS (**a–c**) or Ti6Al4V alloy (**d–f**) untreated (**a, d**) or treated with 0.25 (**b, e**) or 0.5 mM H_2O_2 (**c, f**) 4 h after gel polymerisation. The cells were stained with calcein AM for fluorescent microscopy (scale bar: 300 μm)

angiogenic potential of endothelial cells [98]. In contrast, in another study treatment of microvascular endothelial cells with much higher H_2O_2 concentrations (0.1–0.5 mM for 15 min) induced tubular morphogenesis in type I collagen gel [78]. The differences between the reports could possibly be explained by different cell types and cell culture media and in vitro conditions.

The situation in vivo might be even more complicated, since other cell types can modulate angiogenesis in response to oxidative stress. Thus, in in vitro studies exogenous H_2O_2 has been shown to induce VEGF expression in vascular smooth muscle cells [72], keratinocytes [5] as well as endothelial cells [26]. This was also observed for metal ions. Co^{2+} has been shown to induce VEGF mRNA expression in osteoblasts [81] and VEGF release by fibroblasts [86]. Metal ions can also be involved in the regulation of other growth factors with angiogenic activity. Thus, macrophages exposed to $CoCl_2$ secrete bFGF and PDGF [42]. Metal ions have the ability to modulate angiogenesis directly in vitro. In the same in vitro system as in this study $CoCl_2$ significantly reduced angiogenic potential of endothelial cells [67].

Altogether, it is unclear if oxidative stress on Ti6Al4V alloy has an influence on angiogenesis. Studies on the effect of cathodically polarised Ti6Al4V on the formation of in vitro capillaries by HDMEC could provide an answer to this question.

In conclusion, a permanent state of oxidative stress appears to exist in endothelial cells grown in direct contact with Ti6Al4V surfaces. Although the nature of the stressor is unknown, it is possible that corrosion might take place at the interface of titanium alloy, since ROS formation is expected to occur as a result of the cathodic partial reaction of corrosion.

4.6   Cathodic Half-Reaction of Corrosion as the Possible Source of Oxidative Stress on Ti6Al4V Alloy

Only very few studies have addressed the problem of the cathodic partial reaction of corrosion simulation in vitro to investigate its effects on the cells involved in the implant integration process. It was found in electrochemical studies that oxygen reduction reactions, the main event during cathodic partial reaction of corrosion, can lead to the formation of H_2O_2 on TiO_2 surfaces [8, 9].

The study by Mentus et al. [59] provided evidence that this process is driven by a pH-dependent transition from $2e^-$ to $4e^-$ reduction of oxygen on anodically formed TiO_2 in an electrochemical cell that led to the formation of H_2O_2 and other oxygen radicals as intermediate products. It also showed that oxygen reduction reactions can occur on TiO_2 at any pH value without any particular activation.

To examine the effects of cathodic polarisation products on cellular functions an experimental set-up was developed in which Ti6Al4V discs with cultivated cells were cathodically polarised by applying electrical current to the system [39]. It was demonstrated that endothelial cells grown on Ti6Al4V alloy display the reduction of metabolic activity with increasing current densities applied to Ti6Al4V samples [88]. This was in agreement with lower metabolic activity of HDMEC grown on Ti6Al4V compared to the cells on PS and the further decrease induced by H_2O_2. While lower charge densities (current density per square of metal surface) had no visible effects, higher charge densities induced adverse effects (Fig. 4.7).

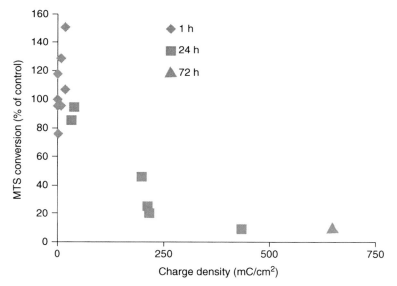

Fig. 4.7   The effect of cathodical partial reaction of corrosion on viability of HDMEC. Metabolic activity of HDMEC on Ti6Al4V was measured with MTS conversion assay after Ti6Al4V cathodical polarisation for 1, 24 or 72. Dependency of MTS conversion in HDMEC on charge density applied to Ti6Al4V (untreated HDMEC are set as 100 %)

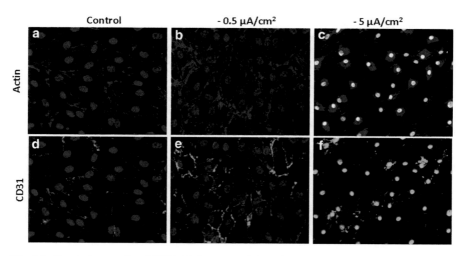

Fig. 4.8 F-actin (**a–c**, *red*) and CD31 (**d–f**, *green*) staining of untreated HDMEC (**a, d**) on Ti6Al4V and after cathodic polarisation of Ti6Al4V alloy for 24 h with −0.5 μA/cm² (**b, e**) and −5 μA/cm² (**c, f**, fluorescent microscopy, nuclei were stained blue with Hoechst 33342, scale bar: 50 μm)

This was also true for cathodic polarisation-induced rearrangements of actin cytoskeleton and loss of intercellular contacts (Fig. 4.8), this being important for the regulation of endothelial permeability during inflammation, at the higher current densities as well as for cell nuclear condensation. Thus, the cells subjected to the products of cathodic half-reaction lost typical peripheral actin ring, as well as intercellular contacts (CD31 staining). This indicated that cathodic half-reaction of corrosion could induce changes in viability and induce a pro-inflammatory phenotype of endothelial cells, thus potentially exerting negative effects on wound healing and stability of an implant.

It was also shown that cathodic polarisation of Ti6Al4V alloy induced drastic elevation in fluorescence of DCF that was given to the cells prior to the treatment (Fig. 4.9). This can result from DCDHF oxidation by H_2O_2 formed during the cathodic partial reaction and freely diffusing through the cell membrane or by ROS induced in the cells by H_2O_2.

Apart from H_2O_2 and ROS formation cathodic half-reaction of corrosion may result in few additional events. First of all hydroxyl ions are formed in oxygen reduction reactions, which might lead to the change in pH. However, the experiments were performed in a buffered system ruling out the possibility of a pH effect. Secondly, oxygen deficiency may occur locally due to oxygen consumption in the cathodic half-reaction. In fact, decreased oxygen concentrations were detected in close proximity to the surface of polarised titanium [24]. This effect correlated with the reduced spreading of osteoblasts on polarised titanium. However, the authors did not take into consideration other events occurring during cathodic half-reactions. Finally, hydrogen might be formed as a result of the cathodic half-reaction. However, this reaction is prevalent at acidic pH or when

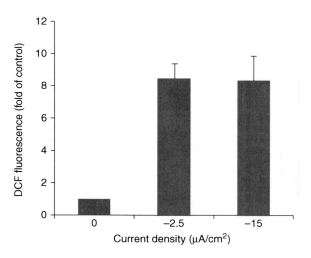

Fig. 4.9 Evaluation of ROS formation in HDMEC grown on Ti6Al4V polarised for 24 h. ROS production was assessed with the DCF assay (DCF fluorescence in untreated cells was set as 1)

higher current densities are applied. This could be excluded, since the current densities used in this study were below oxygen diffusion current density, which is approximately -30 $\mu A/cm^2$ in unstirred air-saturated aqueous electrolytes [80]. Additionally, interactions of the electric field occurring during polarisation with electric properties of cell membranes and electric field-induced changes in the conformation of proteins absorbed to the polarised surface are poorly studied phenomena and can also play their role in the reduction of cell viability in the used model. Electric fields have been shown to direct cell migration during wound healing processes [100].

Interestingly, similar results were obtained in the same model of cathodic polarisation with osteoblastic and macrophage cell lines. Cathodic polarisation of Ti6Al4V alloy induced current density-dependent reduction in cellular metabolic activity that coincided with increased ROS formation and changes in cell morphology. While the reactions of osteoblasts were comparable to H_2O_2 treatment on Ti6Al4V alloy done in parallel, macrophages were more resistant to H_2O_2 [39].

This in vitro model, which allows the exposure of cultivated cells to cathodic corrosion products without distortive effects by anodic reactions/reaction products, offers the possibility to study direct influence of ROS formed at the metal surface on cellular viability and functions. Using current densities that induce production of ROS amounts similar to the in vivo situation could deepen our knowledge about the contribution of the cathodic partial reaction of corrosion to metal-induced adverse tissue reactions. However, the origin and nature of oxidative stress at the metal implant surface has to be considered as a multifactorial process, in which the effects of H_2O_2 and ROS formed on titanium alloy surface may be augmented by ROS derived from inflammatory cells. The possibility of H_2O_2 interaction with the TiO_2 layer, leading to the formation of further ROS, makes the process even more complicated. Upon debris formation damage to the TiO_2 layer occurs, which can amplify the corrosion process. All of these factors point to the contribution of the cathodic half-reaction of corrosion to adverse tissue reactions, leading eventually to aseptic loosening of implants.

4.7   Summary

Aseptic loosening of metal implants is mainly attributed to the formation of metal degradation products. These include particulate debris and corrosion products, such as metal ions (anodic half-reaction) and ROS (cathodic half-reaction). While numerous clinical studies describe various adverse effects of metal degradation products, detailed knowledge of metal-induced cellular reactions, which might be important for possible therapeutic intervention, is not comprehensive. Since endothelial cells are involved in both inflammation and angiogenesis, two processes which are critical for wound healing and integration of metal implants, the effects of different metal alloys and their degradation products on these cells, were investigated. Endothelial cells on Ti6Al4V alloy showed signs of oxidative stress, which was similar to the response of endothelial cells to cathodic partial reaction of corrosion induced directly on Ti6Al4V surfaces. These reactions of endothelial cells to metal degradation products might play their role in the complex processes taking place in the body following metal device implantation. In the worst case this could lead to aseptic loosening of the implant and thus the requirement for revision surgery. Knowledge of molecular mechanisms of metal-induced responses will hopefully provide the possibility to interfere with undesirable processes at the implant/tissue interface, thus extending the lifetime of the implant and the overall success of metal implant applications.

References

1. Bauer TW, Schils J (1999) The pathology of total joint arthroplasty II. Mechanisms of implant failure. Skeletal Radiol 28:483–497
2. Bikondoa O, Pang CL, Ithinin R, Muryn CA, Onishi H, Thornton G (2006) Direct visualization of defect-mediated dissociation of water on $TiO_2(110)$. Nat Mater 5:189–192
3. Blumenthal NC, Cosma V, Jaffe W, Stuchin S (1994) A new technique for quantitation of metal particulates and metal reaction products in tissues near implants. J Appl Biomater 5:191–193
4. Bockris JOM, Reddy AKN, Gamboa-Aldeco M (1998) Modern electrochemistry. Plenum, New York
5. Brauchle M, Funk JO, Kind P, Werner S (1996) Ultraviolet B and H_2O_2 are potent inducers of vascular endothelial growth factor expression in cultured keratinocytes. J Biol Chem 271:21793–21797
6. Cai H (2005) Hydrogen peroxide regulation of endothelial function: origins mechanisms and consequences. Cardiovasc Res 68:26–36
7. Case CP, Langkamer VG, James C, Palmer MR, Kemp AJ, Heap PF, Solomon L (1994) Widespread dissemination of metal debris from implants. J Bone Joint Surg Br 76:701–712

8. Clark T, Johnson D (1997) Activation of titanium electrodes for voltammetric detection of oxygen and hydrogen peroxide in alkaline media. Electroanalysis 9:273–278

9. Clechet P, Martelet C, Martin JR, Olier R (1979) Photoelectrochemical behaviour of TiO_2 and formation of hydrogen-peroxide. Electrochim Acta 24:457–461

10. D'autreaux B, Toledano MB (2007) ROS as signalling molecules: mechanisms that generate specificity in ROS homeostasis. Nat Rev Mol Cell Biol 8:813–824

11. Davies KJ (1987) Protein damage and degradation by oxygen radicals I general aspects. J Biol Chem 262:9895–9901

12. Davies KJ (1999) The broad spectrum of responses to oxidants in proliferating cells: a new paradigm for oxidative stress. IUBMB Life 48:41–47

13. Davies KJ (2000) Oxidative stress antioxidant defenses and damage removal repair and replacement systems. IUBMB Life 50:279–289

14. De Bono DP, Yang WD (1995) Exposure to low concentrations of hydrogen peroxide causes delayed endothelial cell death and inhibits proliferation of surviving cells. Atherosclerosis 114:235–245

15. Dhalla NS, Temsah RM, Netticadan T (2000) Role of oxidative stress in cardiovascular diseases. J Hypertens 18:655–673

16. Dickinson DA, Moellering DR, Iles KE, Patel RP, Levonen AL, Wigley A, Darley-Usmar VM, Forman HJ (2003) Cytoprotection against oxidative stress and the regulation of glutathione synthesis. Biol Chem 384:527–537

17. Diegelmann RF, Evans MC (2004) Wound healing: an overview of acute fibrotic and delayed healing. Front Biosci 9:283–289

18. Disegi JA (2000) Titanium alloys for fracture fixation implants. Injury 31(Suppl 4):14–17

19. Doorn PF, Campbell PA, Amstutz HC (1996) Metal versus polyethylene wear particles in total hip replacements. A review. Clin Orthop Relat Res (329 Suppl): S206–S216

20. Droge W (2002) Free radicals in the physiological control of cell function. Physiol Rev 82:47–95

21. Finkel T (2003) Oxidant signals and oxidative stress. Curr Opin Cell Biol 15:247–254

22. Forman HJ, Torres M (2002) Reactive oxygen species and cell signaling: respiratory burst in macrophage signaling. Am J Respir Crit Care Med 166:S4–S8

23. Galli F, Piroddi M, Annetti C, Aisa C, Floridi E, Floridi A (2005) Oxidative stress and reactive oxygen species. Contrib Nephrol 149:240–260

24. Gilbert JL, Zarka L, Chang E, Thomas CH (1998) The reduction half cell in biomaterials corrosion: oxygen diffusion profiles near and cell response to polarized titanium surfaces. J Biomed Mater Res 42:321–330

25. Gomes A, Fernandes E, Lima JL (2005) Fluorescence probes used for detection of reactive oxygen species. J Biochem Biophys Methods 65:45–80

26. Gonzalez-Pacheco FR, Deudero JJ, Castellanos MC, Castilla MA, Alvarez-Arroyo MV, Yague S, Caramelo C (2006) Mechanisms of endothelial response to oxidative aggression: protective role of autologous VEGF and induction of VEGFR2 by H_2O_2. Am J Physiol Heart Circ Physiol 291:H1395–H1401

27. Goodman SB (1994) The effects of micromotion and particulate materials on tissue differentiation. Bone chamber studies in rabbits. Acta Orthop Scand Suppl 258:1–43

28. Gurtner GC, Werner S, Barrandon Y, Longaker MT (2008) Wound repair and regeneration. Nature 453:314–321

29. Hallab NJ, Jacobs JJ, Skipor A, Black J, Mikecz K, Galante JO (2000) Systemic metal-protein binding associated with total joint replacement arthroplasty. J Biomed Mater Res 49:353–361

30. Harris ED (1992) Regulation of antioxidant enzymes. FASEB J 6:2675–2683

31. Harris WH, Schiller AL, Scholler JM, Freiberg RA, Scott R (1976) Extensive localized bone resorption in the femur following total hip replacement. J Bone Joint Surg Am 58:612–618

32. Holmgren A (1995) Thioredoxin structure and mechanism: conformational changes on oxidation of the active-site sulfhydryls to a disulfide. Structure 3:239–243

33. Jacobs JJ, Gilbert JL, Urban RM (1998) Corrosion of metal orthopaedic implants. J Bone Joint Surg Am 80:268–282
34. Jacobs JJ, Hallab NJ, Skipor AK, Urban RM (2003) Metal degradation products: a cause for concern in metal-metal bearings? Clin Orthop Relat Res (417):139–147
35. Jacobs JJ, Silverton C, Hallab NJ, Skipor AK, Patterson L, Black J, Galante JO (1999) Metal release and excretion from cementless titanium alloy total knee replacements. Clin Orthop Relat Res (358):173–180
36. Jacobs JJ, Skipor AK, Black J, Urban R, Galante JO (1991) Release and excretion of metal in patients who have a total hip-replacement component made of titanium-base alloy. J Bone Joint Surg Am 73:1475–1486
37. Jacobs JJ, Skipor AK, Patterson LM, Hallab NJ, Paprosky WG, Black J, Galante JO (1998) Metal release in patients who have had a primary total hip arthroplasty. A prospective controlled longitudinal study. J Bone Joint Surg Am 80:1447–1458
38. Jefferies H, Coster J, Khalil A, Bot J, Mccauley RD, Hall JC (2003) Glutathione. ANZ J Surg 73:517–522
39. Kalbacova M, Roessler S, Hempel U, Tsaryk R, Peters K, Scharnweber D, Kirkpatrick JC, Dieter P (2007) The effect of electrochemically simulated titanium cathodic corrosion products on ROS production and metabolic activity of osteoblasts and monocytes/macrophages. Biomaterials 28:3263–3272
40. Kinov P, Leithner A, Radl R, Bodo K, Khoschsorur GA, Schauenstein K, Windhager R (2006) Role of free radicals in aseptic loosening of hip arthroplasty. J Orthop Res 24:55–62
41. Konttinen YT, Zhao D, Beklen A, Ma G, Takagi M, Kivela-Rajamaki M, Ashammakhi N, Santavirta S (2005) The microenvironment around total hip replacement prostheses. Clin Orthop Relat Res (430):28–38
42. Kuwabara K, Ogawa S, Matsumoto M, Koga S, Clauss M, Pinsky DJ, Lyn P, Leavy J, Witte L, Joseph-Silverstein J, Al E (1995) Hypoxia-mediated induction of acidic/basic fibroblast growth factor and platelet-derived growth factor in mononuclear phagocytes stimulates growth of hypoxic endothelial cells. Proc Natl Acad Sci U S A 92:4606–4610
43. Larsson J, Persson C, Tengvall P, Lundqvist-Gustafsson H (2004) Anti-inflammatory effects of a titanium-peroxy gel: role of oxygen metabolites and apoptosis. J Biomed Mater Res A 68:448–457
44. Lee MC, Yoshino F, Shoji H, Takahashi S, Todoki K, Shimada S, Kuse-Barouch K (2005) Characterization by electron spin resonance spectroscopy of reactive oxygen species generated by titanium dioxide and hydrogen peroxide. J Dent Res 84:178–182
45. Lee SR, Kwon KS, Kim SR, Rhee SG (1998) Reversible inactivation of protein-tyrosine phosphatase 1B in A431 cells stimulated with epidermal growth factor. J Biol Chem 273: 15366–15372
46. Lee TM, Chang E, Yang CY (2000) A comparison of the surface characteristics and ion release of Ti_6Al_4V and heat-treated Ti_6Al_4V. J Biomed Mater Res 50:499–511
47. Ley K, Laudanna C, Cybulsky MI, Nourshargh S (2007) Getting to the site of inflammation: the leukocyte adhesion cascade updated. Nat Rev Immunol 7:678–689
48. Li J, Zhang YP, Kirsner RS (2003) Angiogenesis in wound repair: angiogenic growth factors and the extracellular matrix. Microsc Res Tech 60:107–114
49. Lin HY, Bumgardner JD (2004) In vitro biocorrosion of Ti-6Al-4V implant alloy by a mouse macrophage cell line. J Biomed Mater Res A 68:717–724
50. Long M, Rack HJ (1998) Titanium alloys in total joint replacement–a materials science perspective. Biomaterials 19:1621–1639
51. Lu SC (2000) Regulation of glutathione synthesis. Curr Top Cell Regul 36:95–116
52. Lum H, Roebuck KA (2001) Oxidant stress and endothelial cell dysfunction. Am J Physiol Cell Physiol 280:C719–C741
53. Macdonald SJ (2004) Metal-on-metal total hip arthroplasty: the concerns. Clin Orthop Relat Res (429):86–93

54. Malhotra JD, Miao H, Zhang K, Wolfson A, Pennathur S, Pipe SW, Kaufman RJ (2008) Antioxidants reduce endoplasmic reticulum stress and improve protein secretion. Proc Natl Acad Sci U S A 105:18525–18530

55. Mandelin J, Li TF, Liljestrom M, Kroon ME, Hanemaaijer R, Santavirta S, Konttinen YT (2003) Imbalance of RANKL/RANK/OPG system in interface tissue in loosening of total hip replacement. J Bone Joint Surg Br 85:1196–1201

56. Marti A (2000) Cobalt-base alloys used in bone surgery. Injury 31(Suppl 4):18–21

57. Martin P, Leibovich SJ (2005) Inflammatory cells during wound repair: the good, the bad, and the ugly. Trends Cell Biol 15:599–607

58. Meilhac O, Zhou M, Santanam N, Parthasarathy S (2000) Lipid peroxides induce expression of catalase in cultured vascular cells. J Lipid Res 41:1205–1213

59. Mentus SV (2004) Oxygen reduction on anodically formed titanium dioxide. Electrochim Acta 50:27–32

60. Midwood KS, Williams LV, Schwarzbauer JE (2004) Tissue repair and the dynamics of the extracellular matrix. Int J Biochem Cell Biol 36:1031–1037

61. Milosev I, Metikos-Hukovic M, Strehblow HH (2000) Passive film on orthopaedic TiAlV alloy formed in physiological solution investigated by X-ray photoelectron spectroscopy. Biomaterials 21:2103–2113

62. Mu Y, Kobayashi T, Sumita M, Yamamoto A, Hanawa T (2000) Metal ion release from titanium with active oxygen species generated by rat macrophages in vitro. J Biomed Mater Res 49:238–243

63. Murphy RC, Johnson KM (2008) Cholesterol reactive oxygen species and the formation of biologically active mediators. J Biol Chem 283:15521–15525

64. Ni M, Lee AS (2007) ER chaperones in mammalian development and human diseases. FEBS Lett 581:3641–3651

65. Ozmen I, Naziroglu M, Okutan R (2005) Comparative study of antioxidant enzymes in tissues surrounding implant in rabbits. Cell Biochem Funct 24:275–281

66. Pan J, Liao H, Leygraf C, Thierry D, Li J (1998) Variation of oxide films on titanium induced by osteoblast-like cell culture and the influence of an H_2O_2 pretreatment. J Biomed Mater Res 40:244–256

67. Peters K, Schmidt H, Unger RE, Otto M, Kamp G, Kirkpatrick CJ (2002) Software-supported image quantification of angiogenesis in an in vitro culture system: application to studies of biocompatibility. Biomaterials 23:3413–3419

68. Plant SD, Grant DM, Leach L (2005) Behaviour of human endothelial cells on surface modified NiTi alloy. Biomaterials 26:5359–5367

69. Pohler OE (2000) Unalloyed titanium for implants in bone surgery. Injury 31(Suppl 4):7–13

70. Rahman I, Bel A, Mulier B, Lawson MF, Harrison DJ, Macnee W, Smith CA (1996) Transcriptional regulation of gamma-glutamylcysteine synthetase-heavy subunit by oxidants in human alveolar epithelial cells. Biochem Biophys Res Commun 229:832–837

71. Rhee SG, Chang TS, Bae YS, Lee SR, Kang SW (2003) Cellular regulation by hydrogen peroxide. J Am Soc Nephrol 14:S211–S215

72. Ruef J, Hu ZY, Yin LY, Wu Y, Hanson SR, Kelly AB, Harker LA, Rao GN, Runge MS, Patterson C (1997) Induction of vascular endothelial growth factor in balloon-injured baboon arteries. A novel role for reactive oxygen species in atherosclerosis. Circ Res 81:24–33

73. Sabokbar A, Pandey R, Quinn JM, Athanasou NA (1998) Osteoclastic differentiation by mononuclear phagocytes containing biomaterial particles. Arch Orthop Trauma Surg 117:136–140

74. Schraufstatter IU, Hinshaw DB, Hyslop PA, Spragg RG, Cochrane CG (1986) Oxidant injury of cells DNA strand-breaks activate polyadenosine diphosphate-ribose polymerase and lead to depletion of nicotinamide adenine dinucleotide. J Clin Invest 77:1312–1320

75. Schroder M, Kaufman RJ (2005) ER stress and the unfolded protein response. Mutat Res 569:29–63

76. Senderowicz AM (2003) Small-molecule cyclin-dependent kinase modulators. Oncogene 22:6609–6620

77. Shiraiwa M, Goto T, Yoshinari M, Koyano K, Tanaka T (2002) A study of the initial attachment and subsequent behavior of rat oral epithelial cells cultured on titanium. J Periodontol 73:852–860
78. Shono T, Ono M, Izumi H, Jimi SI, Matsushima K, Okamoto T, Kohno K, Kuwano M (1996) Involvement of the transcription factor NF-kappaB in tubular morphogenesis of human microvascular endothelial cells by oxidative stress. Mol Cell Biol 16:4231–4239
79. Sinha RK, Morris F, Shah SA, Tuan RS (1994) Surface composition of orthopaedic implant metals regulates cell attachment spreading and cytoskeletal organization of primary human osteoblasts in vitro. Clin Orthop Relat Res (305):258–272
80. Song FM, Kirk DW, Graydon JW, Cormack DE (2002) CO_2 corrosion of bare steel under an aqueous boundary layer with oxygen. J Electrochem Soc 149:479–486
81. Steinbrech DS, Mehrara BJ, Saadeh PB, Greenwald JA, Spector JA, Gittes GK, Longaker MT (2000) VEGF expression in an osteoblast-like cell line is regulated by a hypoxia response mechanism. Am J Physiol Cell Physiol 278:C853–C860
82. Sundfeldt M, Carlsson LV, Johansson CB, Thomsen P, Gretzer C (2006) Aseptic loosening not only a question of wear: a review of different theories. Acta Orthop 77:177–197
83. Taylor GC, Waddington RJ, Moseley R, Williams KR, Embery G (1996) Influence of titanium oxide and titanium peroxy gel on the breakdown of hyaluronan by reactive oxygen species. Biomaterials 17:1313–1319
84. Tengvall P, Elwing H, Sjoqvist L, Lundstrom I, Bjursten LM (1989) Interaction between hydrogen peroxide and titanium: a possible role in the biocompatibility of titanium. Biomaterials 10:118–120
85. Tengvall P, Lundstrom I, Sjoqvist L, Elwing H, Bjursten LM (1989) Titanium-hydrogen peroxide interaction: model studies of the influence of the inflammatory response on titanium implants. Biomaterials 10:166–175
86. Trompezinski S, Pernet I, Mayoux C, Schmitt D, Viac J (2000) Transforming growth factor-beta1 and ultraviolet A1 radiation increase production of vascular endothelial growth factor but not endothelin-1 in human dermal fibroblasts. Br J Dermatol 143:539–545
87. Tsaryk R, Kalbacova M, Hempel U, Scharnweber D, Unger RE, Dieter P, Kirkpatrick CJ, Peters K (2007) Response of human endothelial cells to oxidative stress on Ti_6Al_4V alloy. Biomaterials 28:806–813
88. Tsaryk R, Peters K, Unger RE, Scharnweber D, Kirkpatrick CJ (2007) The effects of metal implants on inflammatory and healing processes. Int J Mat Res 98:622–629
89. Tucci M, Baker R, Benghuzzi H, Hughes J (2000) Levels of hydrogen peroxide in tissues adjacent to failing implantable devices may play an active role in cytokine production. Biomed Sci Instrum 36:215–220
90. Ushio-Fukai M, Alexander RW (2004) Reactive oxygen species as mediators of angiogenesis signaling: role of NAD(P)H oxidase. Mol Cell Biochem 264:85–97
91. Valko M, Rhodes CJ, Moncol J, Izakovic M, Mazur M (2006) Free radicals metals and antioxidants in oxidative stress-induced cancer. Chem Biol Interact 160:1–40
92. Veal EA, Day AM, Morgan BA (2007) Hydrogen peroxide sensing and signaling. Mol Cell 26:1–14
93. Werner S, Grose R (2003) Regulation of wound healing by growth factors and cytokines. Physiol Rev 83:835–870
94. Williams DF (1981) Titanium and titanium alloys. In: Williams DF (ed) Biocompatibility of clinical implant materials. CRC, Boca Raton, FL
95. Winterbourn CC (2008) Reconciling the chemistry and biology of reactive oxygen species. Nat Chem Biol 4:278–286
96. Wooley PH, Schwarz EM (2004) Aseptic loosening. Gene Ther 11:402–407
97. Wu G, Fang YZ, Yang S, Lupton JR, Turner ND (2004) Glutathione metabolism and its implications for health. J Nutr 134:489–492
98. Yasuda M, Ohzeki Y, Shimizu S, Naito S, Ohtsuru A, Yamamoto T, Kuroiwa Y (1999) Stimulation of in vitro angiogenesis by hydrogen peroxide and the relation with ETS-1 in endothelial cells. Life Sci 64:249–258

99. Zhang Z, Huang C, Li J, Leonard SS, Lanciotti R, Butterworth L, Shi X (2001) Vanadate-induced cell growth regulation and the role of reactive oxygen species. Arch Biochem Biophys 392:311–320

100. Zhao M, Song B, Pu J, Wada T, Reid B, Tai G, Wang F, Guo A, Walczysko P, Gu Y, Sasaki T, Suzuki A, Forrester JV, Bourne HR, Devreotes PN, Mccaig CD, Penninger JM (2006) Electrical signals control wound healing through phosphatidylinositol-3-OH kinase-gamma and PTEN. Nature 442:457–460

101. Zumdahl S (2007) Chemical principles. Brooks Cole, http://www.amazon.com/Chemical-Principles-Steven-S-Zumdahl/dp/061894690X

Chapter 5
Comparative Properties of Ethyl, *n*-Butyl, and *n*-Octyl Cyanoacrylate Bioadhesives Intended for Wound Closure

Ana María Villarreal-Gómez, Rafael Torregrosa-Coque, and José Miguel Martín-Martínez

5.1 Introduction

Adhesives intended for use in the bonding of human tissues can be called bioadhesives. One of the advantages of using bioadhesives to join human tissues as compared to the traditional suture is the creation of homogeneous and uniform distribution of stresses all along the joint. Furthermore, the bioadhesives are easy to apply, they are less traumatic to patient, their use reduces the surgical time, avoids the mechanical damage produced in the tissues by suture, inhibits scare formation, and there is no need to remove the remaining sutures or bandages after surgical practice.

Cyanoacrylate monomer polymerizes mainly in the presence of bases, such as water. Skin has a high concentration of water, so cyanoacrylate monomers can be used as adhesives for surgical practice and wound closure. Cyanoacrylate monomers also produce strong and quick adhesion in wound closure.

Since 1950 cyanoacrylate monomers have been used as adhesives [1, 2]. Several recent studies [3–12] have shown the effectiveness of different bioadhesives as an alternative to suture practice. Particularly, cyanoacrylate adhesives have shown excellent performance as skin adhesives, surgical glues, and embolitic materials [4–7], in skin closure, in plastic surgery, or in osteosynthesis (ethyl cyanoacrylate (ECN) polymerized by ultrasounds) [8, 9], in the therapeutic embolization of cerebral arteriovenous malformations, in gastric variceal bleeding or corneal perforations [10, 11], as well as in ocular strabismus surgery [12]. Cañizares-Grupera

A.M. Villarreal-Gómez
Bioadhesives Medtech Solutions, Elche, Alicante 03206, Spain
e-mail: ana.villarreal@adhbio.com

R. Torregrosa-Coque • J.M. Martín-Martínez (✉)
Adhesion and Adhesives Laboratory, University of Alicante, Alicante 03080, Spain
e-mail: rafael.torregrosa@ua.es; jm.martin@ua.es

I. Antoniac (ed.), *Biologically Responsive Biomaterials for Tissue Engineering*,
Springer Series in Biomaterials Science and Engineering 1,
DOI 10.1007/978-1-4614-4328-5_5, © Springer Science+Business Media New York 2013

et al. [13] used alkyl cyanoacrylates as skin sealants for plastic surgery. The study compares the effectiveness of the suture versus bioadhesives in 100 patients with 225 skin wounds and 16 nasal implants. Similar performance was obtained by using the suture and the bioadhesive. More recently poly (alkyl cyanoacrylates) have been proposed as raw materials for the synthesis of nanoparticles intended as drug delivery carriers [14].

If medical use of cyanoacrylate glues goes back to the half of the last century, especially for military applications, only in the last two decades the US Food and Drug Administration (FDA) approved some classes of cyanoacrylates as biocompatible compounds, http://www.fda.gov/AboutFDA/CentersOffices/CDRH/CDRHReports/ucm127557.htm. With the recent decision of an FDA expert panel of reclassifying the cyanoacrylate topical skin adhesives from class III device (that requires performance of new clinical trial to receive approval) to class II device (requiring only the demonstration of substantial equivalency to a currently approved predicate device), it can be anticipated that many new cyanoacrylate monomers will become available in the United States in the near future.

However, despite producing effective sealing, the use of cyanoacrylate monomers showed exothermal cure and high stiff polymerized product of considerable strength but fragile at the same time. In addition, erosion, ulceration, and areas of necrosis in surrounding (or adjacent) tissues may occur. One easy route to reduce the exothermic reaction and stiffness of the cured cyanoacrylates is to increase the length of the alkyl hydrocarbon chain in the monomer.

To the best of our knowledge there are only scarce papers and literature comparing the performance of cyanoacrylate adhesives with different hydrocarbon chain length and most of them are devoted to clinical studies. Recently, Dossi et al. [3] studied the anionic polymerization of ethyl, n-butyl, and n-octyl cyanoacrylate in water. They found that an aqueous dispersion medium at pH around 5 and 65 °C gave the best cure performance and polymer properties. Both low- and high-molecular-weight polymeric chains were found but at room temperature the amount of high-molecular-weight polymer became more noticeable. Charters [15] compared three tissue commercial adhesives: *Indermil®* (ECN), *Liquiband®* (n-butyl cyanoacrylate (BCN)), and *Dermabond®* (octyl cyanoacrylate) in the treatment of 39,000 children patients. None of the glues were reported to be completely pain free. However, the *Liquiband®* tissue adhesive produced an average pain score of only 0.1 as compared to the *Dermabond®* tissue adhesive scored as 0.97. *Liquiband®* was the best tissue adhesive in terms of wound closure and ease of use. However, the only tissue adhesive with complete success was *Indermil®*. All of the tissue adhesives examined produced satisfactory results in terms of wound closure. N. G. Senchenya et al. [16] polymerized different cyanoacrylate monomers by thermal treatment and established that their glass transition temperatures depended on both the chemical structure and the kinetics of curing of the monomers. On the other hand, they showed that the elastic modulus decreased by increasing the length of the alkyl hydrocarbon chain of the cyanoacrylate monomer. D. H. Park et al. [17] studied the kinetics of degradation and cytotoxicity of ethyl, 2-octyl, *n*-octyl, and ethylhexyl cyanoacrylates by using different polymerization procedures. They showed that the cyanoacrylate powders degraded faster than the films and that the ECN gave the faster degradation and higher level of cytotoxicity (mainly due to formaldehyde evolution).

Considering these previous findings in the existing literature, the aim of this study was to compare the physical-chemical properties of several cyanoacrylate monomers and polymers having linear alkyl hydrocarbon chains with different length. In this study the preliminary experimental results obtained up to now in our laboratory are reported as the study is currently under development.

5.2 Experimental

5.2.1 Materials

Three cyanoacrylate monomers were used in this study (Fig. 5.1): ECN (R=CH_2–CH_3), BCN (R=$(CH_2)_3$–CH_3), and (OCN (R=$(CH_2)_7$–CH_3).

The ECN monomer was prepared in our laboratory. In the first step (Fig. 5.2) polyethyl cyanoacrylate was obtained by reacting ethyl cyanoacetate and formaldehyde (both CP grade and provided by Aldrich, Barcelona, Spain) in the presence of a base catalyst—i.e., piperidinium chloride (Knoevenagel reaction). The catalyst produced hydrogen abstraction in α position to the carboxyl group in the ethyl cyanoacetate, and the resulting carbanion was added to formaldehyde, giving the condensation product. The monomer was produced by depolymerization of the polyethyl cyanoacrylate by direct heating in a burner [18]. The resulting ECN monomer was further purified through low-pressure distillation. The ECN obtained had an Ubbelohde viscosity of 2.21 cStokes.

Both the *n*-butyl and *n*-octyl cyanoacrylate monomers used in this study were commercial products. BCN monomer was *Vetbond*® manufactured by 3M (St. Paul, Minnesota, USA) and contains hydroquinone stabilizer and blue dye. OCN was *Dermabond*® manufactured by Ethicon (Somerville, New Jersey, USA) and contains thickening agent, stabilizer, and violet dye.

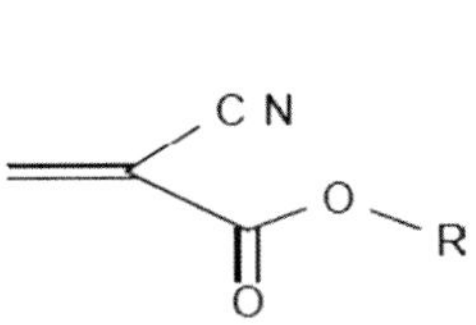

R = CH_2-CH_3 ethyl cyanoacrylate

R = $(CH_2)_3$-CH_3 n-butyl cyanoacrylate

R = $(CH_2)_7$-CH_3 n-octyl cyanoacrylate

Fig. 5.1 Cyanoacrylate monomers used in this study

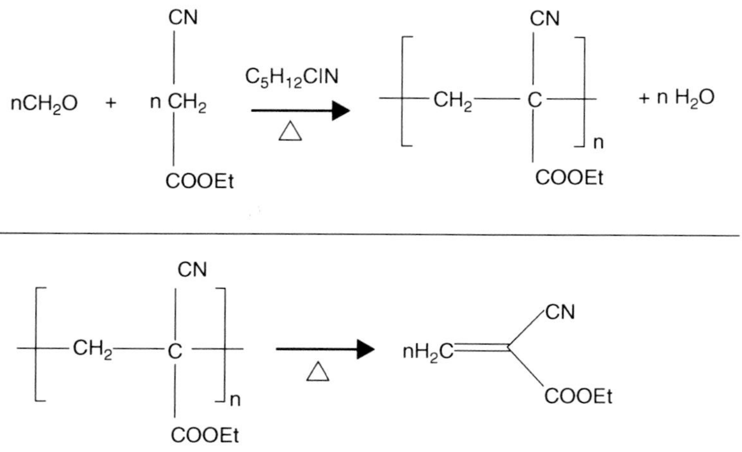

Fig. 5.2 Synthesis of ethyl cyanoacrylate monomer (Et: CH_2–CH_3)

5.2.1.1 Polymerization of the Cyanoacrylate Monomers

Cyanoacrylate polymer films were prepared by adding water to the corresponding mono-mers. 2 ml of cyanoacrylate monomer and 2 ml of distilled water (1:1 vol/vol) were mixed in an polyethylene container under ambient conditions. Although the monomers began to polymerize quickly when they came into contact with water, the polymerization reactions lasted for 1 week. After taking out the polymer films from the remaining water, they were maintained at room temperature for 24 h before characterization.

5.2.2 Experimental Techniques

5.2.2.1 Infrared Spectroscopy

The ATR-IR spectra of the cyanoacrylates were obtained in Tensor 27 Bruker spec-trometer (Bruker Optik GmbH, Madrid, Spain). The attenuated total multiple reflection method was employed (ATR-IR), and a diamond prism (Golden gate device) was used; the incident angle of the IR beam was 45° and 60 scans were obtained and averaged at a resolution of 4 cm^{-1}.

5.2.2.2 ^{1}H and ^{13}C Nuclear Magnetic Resonance

Nuclear Magnetic Resonance (NMR) experiments were carried out in Bruker AC-400 spectrometer (Bruker, Rheinstetten. Germany), provided with a magnet of 400 MHz, dual probe ^{1}H/^{13}C (5 mm), variable temperature unit, and automated sample exchanger up to 60 samples. The experiments were carried out in special tubes containing 30–40 mg sample, deuterated chloroform ($CDCl_3$) as solvent, and tetramethylsilane (TMS) as standard. The data were processed by using Mestrec 5.0. software (Mestrenova 2007) (Mestrelab Research S. L., Santiago de Compostela, Spain).

5.2.2.3   Thermal Gravimetric Analysis

Thermal gravimetric analysis (TGA) studies were carried out in TA TGA Q500 instrument (TA instruments, New Castle, Delaware, USA) under nitrogen atmosphere (flow rate: 100 ml/min). Samples of cyanoacrylate (10–15 mg) were heated from room temperature up to 400 °C by using a heating rate of 10 °C/min.

5.2.2.4   Scanning Electron Microscopy

The topography of the polymerized cyanoacrylates was analyzed by scanning electron microscopy (SEM). Hitachi S-3000N microscope (Tokyo, Japan) was used.

5.2.2.5   Single Lap-Shear Tests

Single lap-shear tests of aluminum 5754/cyanoacrylate adhesive/aluminum 5754 and pig skin/cyanoacrylate adhesive/pig skin joints were carried out to measure the adhesion properties of the cyanoacrylates.

Aluminum 5754/cyanoacrylate adhesive/aluminum 5754 joints were used to determine the intrinsic adhesion properties of the polymerized cyanoacrylates. Before joint formation, the aluminum was smoothly roughened with Scotch Brite scourer and later wiped with 2-propanol. Fifteen minutes later, 0.05 ml of cyanoacrylate monomer was placed on each aluminum test piece and immediately joined without applying pressure in a square contact area of 9 cm^2 (3×3 cm) (Fig. 5.3). Single lap-shear tests were carried out 72 h after joint formation in Instron 8516 universal testing machine (Instron Ltd., Buckinghamshire, UK) by using a pulling rate of 100 mm/min. Five replicates were measured and averaged with an error lower than ±0.1 MPa.

Pig skin/cyanoacrylate adhesive/pig skin joints were carried out to determine the immediate adhesion of the cyanoacrylate monomers. Pig skin was used to simulate the joints that potentially can be produced in wound closure practice. Pig skin test samples of dimensions 60×30 mm were cut from the upper leg pieces of freshly sacrificed pig (kindly supplied by Juan Carlos Lillo Garrigós' butcher shop, San Vicente del Raspeig, Alicante, Spain). The pig skin test samples were immersed for 5 min in 60 ml physiological serum solution (Fleboplast, Grifols, Barcelona, Spain) to impart homogeneous and controlled humidity. 10 s after removal of the test pieces from the solution, they were surface dried with filter paper for 10 s and the test samples were maintained under ambient conditions for 30 s.

Fig. 5.3 Aluminum 5754/
cyanoacrylate monomer/
aluminum 5754 adhesive
joints used in single lap-shear
tests

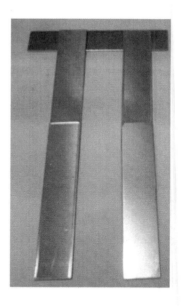

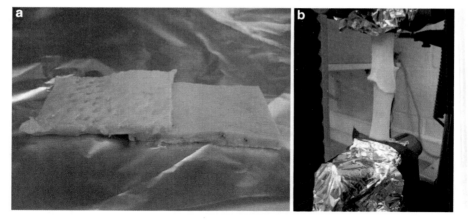

Fig. 5.4 (**a**) Pig skin/cyanoacrylate adhesive/pig skin joint used in single lap-shear tests.
(**b**) Single lap-shear test of pig skin/cyanoacrylate adhesive/pig skin joint

   0.05 ml of cyanoacrylate monomer was placed by means of a syringe on each pig
skin test piece and the liquid monomer was spread over the bonding area by means
of the syringe needle. The two test pieces were immediately placed in contact and
pressed by means of two fingers for 30 s (Fig. 5.4a). 60 s later, the single lap-shear
test was carried out in Instron 4411 universal testing machine (Instron Ltd.,
Buckinghamshire, UK) by using a pulling rate of 100 mm/min (Fig. 5.4b). Five
replicates were measured and averaged with an error of ±0.2 kPa.

5.3   Results and Discussion

5.3.1   Characterization of the Cyanoacrylate Monomers

Some physical-chemical properties of the cyanoacrylate monomers used in this study are given in Table 5.1. As expected, the viscosity and boiling point of the monomers increase by increasing the length of the alkyl hydrocarbon chain.

Figure 5.5 shows the IR spectra of the cyanoacrylate monomers and the assignment of the main IR bands is given in Table 5.2.

The main characteristic bands correspond to=CH$_2$ (3,128 cm^{-1}), C≡N (2,238 cm^{-1}), C=O (1,731 cm^{-1}), C=C (963, 1,614 cm^{-1}), and C–O–C (1,157, 1,170–1,189, 1,286 cm^{-1}) groups. By increasing the length of the alkyl chain in the cyanoacrylate an increase in the relative intensity of the CH$_2$ and CH$_3$ stretching bands is produced.

On the other hand, the relative intensity of the C–O–C stretching bands decreases by increasing the length of the hydrocarbon chain but the bands at 1,170–1,189 cm^{-1} are better distinguished.

Table 5.1 Some properties of the cyanoacrylate monomers

Alkyl group	Molecular weight (g/mol)	Boiling point (°C/Torr)	Density (20 °C) (g/ml)	Viscosity (25 °C) (mPa s)
Ethyl	125	60/3	1.040	1.86
n-Butyl	153	68/1.8	0.989	2.07
n-Octyl	209	117/1.8	0.931	3.92

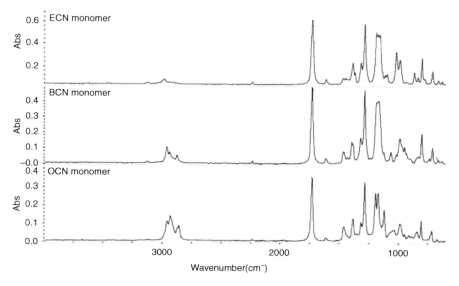

Fig. 5.5 ATR-IR spectra of the cyanoacrylate monomers

Table 5.2 Assignment of the main IR bands (Fig. 5.5) of the cyanoacrylate monomers

Wavenumber (cm^{-1})	Assignment
3,128	=CH$_2$ st.
2,860–2,984	–CH$_2$ st., –CH$_3$ st.
2,238	–C≡N st.
1,731	–C=O st.
1,614	C=C st.
1,465	–CH$_3$ as st.
1,387	–CH$_3$ sim st.
1,447	–CH$_2$ bending
1,157–1,286	C–O–C st.
963	C=C bending
803	C=C–C–O st.
714	(CH$_2$)$_n$ $n \geq 4$

st. Stretching

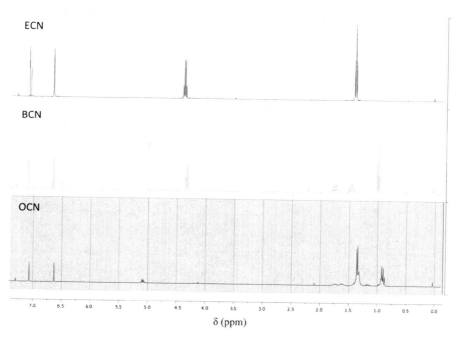

Fig. 5.6 ^{1}H NMR spectra of the cyanoacrylate monomers

NMR spectroscopy was used for more accurate assessment of the chemical structure of the cyanoacrylate monomers. The ^{1}H NMR spectra (Fig. 5.6) show the typical signals of the protons in the cyanoacrylate monomers.

Table 5.3 shows as typical example the assignment of the protons of BCN monomer. The intensity of the signals corresponding to the protons of the C=C bond in

Table 5.3 Assignment of the main characteristic signals in the ¹H NMR spectrum of the *n*-butyl cyanoacrylate monomer

H	δ(ppm)	Multiplicity	Integral
H_1	7.06	Single	1.00
H_2	6.62	Single	1.00
H_3	4.29	Triple	2.01
H_4	1.70	Multiple	2.20
H_5	1.38	Multiple	2.06
H_6	0.96	Triple	3.00

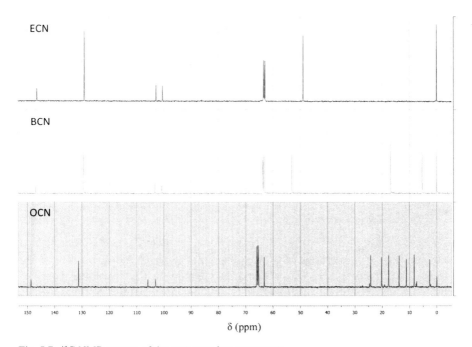

Fig. 5.7 ¹³C NMR spectra of the cyanoacrylate monomers

the cyanoacrylate monomer decreases by increasing the length of the alkyl chain. The protons of the CH_2 groups appear at 4.3 ppm and the signal is less marked by increasing the length of the alkyl chain in the cyanoacrylate monomer; however, the signals of the protons at the end of the alkyl chain (below 1 ppm) are more intense in the OCN. The signals at 2.18 and 1.25 ppm in the ¹H NMR spectrum of the BCN monomer should correspond to aliphatic protons but they cannot be assigned to its chemical structure; these signals can be ascribed to additives in the commercial product. Similarly, the signals of the CH_2 of the OCN do not appear in its ¹H NMR spectrum but several overlapped signals are shown at 0.95 and 1.42 ppm that correspond to protons of this compound and to additives in the commercial product.

The ¹³C NMR spectra of the cyanoacrylate monomers (Fig. 5.7) show the typical signals of the different carbons in their structure. Table 5.4 shows as typical example

Table 5.4 Assignment of the
most characteristic signals in
the ^{13}C NMR spectrum of the
n-butyl   cyanoacrylate
monomer

C	δ (ppm)
C_1	129.38
C_2	103.25
C_3	100.83
C_4	146.93
C_5	63.46
C_6	53.11
C_7	16.93
C_8	5.39

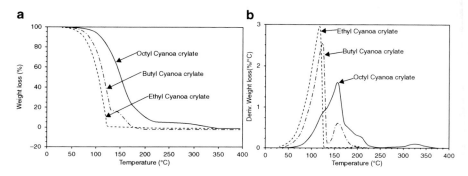

Fig. 5.8 (**a**) TGA thermograms of the cyanoacrylate monomers. (**b**) Derivative curves of the TGA thermograms of the cyanoacrylate monomers

the assignment of the carbons of BCN monomer. The signals corresponding to the carbons corresponding to C=C, C=O, and C≡N groups displace to higher values by increasing the length of the alkyl hydrocarbon chain in the cyanoacrylate monomer. The ^{13}C NMR spectrum of the OCN shows a more noticeable content of aliphatic carbons with respect to the signals of the other cyanoacrylate monomers.

In this study TGA was found useful and sensitive to analyze the structure of the cyanoacrylate monomers. Figure 5.8a shows the TGA thermograms of the three cyanoacrylate monomers and the corresponding derivative curves are given in Fig. 5.8b; the derivative curves are useful for showing more precisely the thermal decompositions in the cyanoacrylate monomers.

The TGA curves of the three cyanoacrylate monomers show the first decomposition at about 120 °C (Table 5.5), likely due to formaldehyde evolution. This first decomposition becomes less marked by increasing the length of the alkyl chain of the cyanoacrylate monomer, i.e., the weight loss at this temperature decreases from ethyl to n-octyl cyanoacrylate monomer. On the other hand, the thermal stability of the cyanoacrylate monomer increases by increasing the length of the alkyl hydrocarbon chain, likely due to the increase in its molecular weight.

The presence of more than one decomposition step in the cyanoacrylate monomers is an indication of the presence of oligomers in their structure. Whereas the ECN is mainly composed of monomer (98.7 wt% is lost at 117 °C), the BCN

Table 5.5 Weight loss and temperature of the thermal decompositions of the cyanoacrylate monomers

Monomer	T1 (°C)	Weight loss 1 (wt%)	T2 (°C)	Weight loss 2 (wt%)	T3 (°C)	Weight loss 3 (wt%)	T4 (°C)	Weight loss 4 (wt%)	T5 (°C)	Weight loss 5 (wt%)
ECN	117	98.7	–	–	–	–	196	1.3	–	–
BCN	124	83.9	157	17.8	–	–	–	–	–	–
OCN	126	25.3	156	48.2	175	13.5	208	7.1	327	5.9

TGA experiments

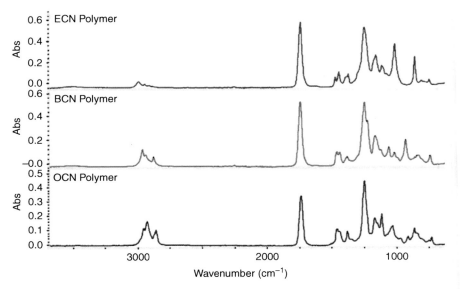

Fig. 5.9 ATR-IR spectra of the cyanoacrylate polymers

contains monomer and some dimmers (17.8 wt% is lost at 157 °C), and the OCN has only a minor content of monomer (only 25.3 wt% is lost at 126 °C) and shows four additional decompositions (Fig 5.8a, b; Table 5.5).

These results are in agreement with recent literature [3] that also described the existence of different decompositions in the TGA thermograms of BCN polymer ascribed to the existence of different structures in the polymer.

5.3.2   Characterization of the Cyanoacrylate Polymers

Cyanoacrylate monomers were polymerized at room temperature by adding water. The properties of the resulting cyanoacrylate polymers were characterized and compared with those of the monomers to establish the changes produced during polymerization.

Figure 5.9 shows the ATR-IR spectra of the cyanoacrylate polymers.

In general, the assignment of the main IR bands is similar to that of the monomers (Table 5.2), although the relative intensity of the bands is different and two new distinct intense bands at 856 and 1,240 cm^{-1} due to new $C-O-C$ groups appear after polymerization of the cyanoacrylate monomers. As expected, the intensity of the CH_2 group decreases by increasing the length of the alkyl hydrocarbon chain of the cyanoacrylate and because of the polymerization the intensity of the band at 2,239 cm^{-1} due to the cyanoacrylate group is strongly reduced.

Figure 5.10 compares as typical example two significant regions of the ATR-IR spectra of the ECN monomer and polymer. The polymerization produces the disappearance of the C=C bands at 1,615 and 3,128 cm^{-1} as the polymerization of the

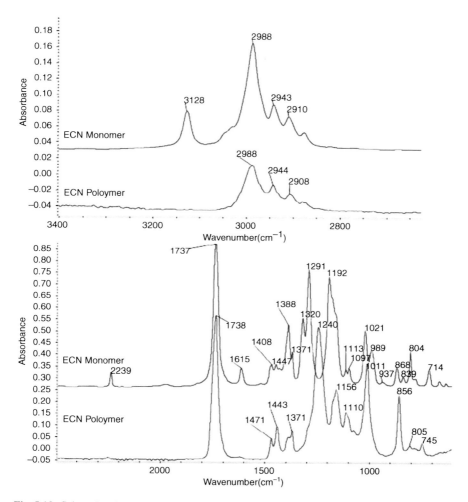

Fig. 5.10 Selected regions in the ATR-IR spectra of the ethyl cyanoacrylate monomer and polymer

cyanoacrylate is produced throughout these C=C bonds [1]. Furthermore, after polymerization of the ECN the intensity of the CH_3 group decreases.

The thermal properties of the cyanoacrylate polymers also differ from those of the corresponding monomers. Figure 5.11a shows as typical example the TGA thermograms of the BCN monomer and polymer. After polymerization the degradation temperature of the cyanoacrylate increases noticeably and the percentages of the two different structures found in the TGA of the monomer (Fig. 5.8a) change and almost similar content is obtained. The different polymer structures can be better distinguished in the derivative curves of the TGA thermograms given in Fig. 5.11b.

Table 5.6 summarizes the weight loss and temperature of the different thermal decompositions of the ethyl, *n*-butyl, and *n*-octyl cyanoacrylate polymers. In general,

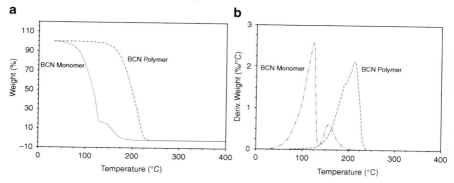

Fig. 5.11 (**a**) TGA thermograms of *n*-butyl cyanoacrylate monomer and polymer. (**b**) Derivative curves of the TGA thermograms of *n*-butyl cyanoacrylate monomer and polymer

Table 5.6 Weight loss and temperature of the different decompositions of the cyanoacrylate polymers

Polymer	T1 (°C)	Weight loss 1 (wt%)	T2 (°C)	Weight loss 2 (wt%)
ECN	196	99.1	–	–
BCN	191	42.8	214	57.2
OCN	231	94.6	344	5.4

TGA experiments

the temperature at which decomposition is produced increases by increasing the length of the alkyl hydrocarbon chain of the cyanoacrylate. Whereas the ECN polymer shows a unique decomposition at 199 °C, both the *n*-butyl and the *n*-octyl cyanoacrylate polymers show two thermal decompositions corresponding to two different structures.

SEM was used to analyze the topography of the cyanoacrylate polymers. Figure 5.12 shows the SEM micrographs of the three polymers.

Whereas the ECN polymer shows a homogeneous and smooth surface, the *n*-butyl and, more markedly, the *n*-octyl cyanoacrylate polymers show a heterogeneous topography consisting in shredded polymer chains. This indicates a different extent of polymerization and the increase in the length of the alkyl hydrocarbon chain cyanoacrylate produces less cohesive structure. This is in agreement with the decrease in elastic modulus of the cyanoacrylate polymer by increasing the length of the alkyl hydrocarbon chain reported elsewhere [16]. To the best of our knowledge, this is the first time in the literature in which the differences in topography in the polymerized cyanoacrylates are shown.

5.3.3 Adhesion Properties of Cyanoacrylate Adhesives

Adhesion properties of the cyanoacrylate monomers were measured by means of single lap-shear tests. Aluminum/cyanoacrylate adhesive and pig skin/cyanoacrylate adhesive joints were prepared to measure the adhesion of the bioadhesives once

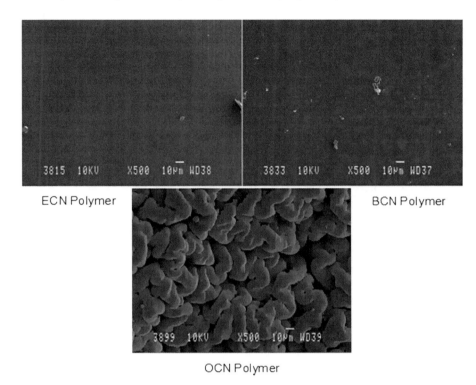

ECN Polymer   BCN Polymer

OCN Polymer

Fig. 5.12   SEM micrographs of the cyanoacrylate polymers

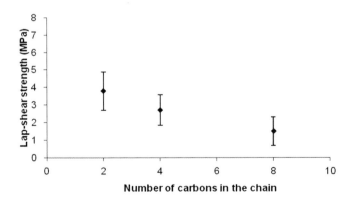

Fig. 5.13   Single lap-shear strength values of aluminum 5754/cyanoacrylate monomer/aluminum 5754 adhesive joints. Values obtained 72 h after joint formation

they are cured and their performance in wound closure, i.e., just after application, respectively.

Figure 5.13 shows the single lap-shear strength values of aluminum 5754/cyanoacrylate adhesive joints. The single lap-shear strength value decreases

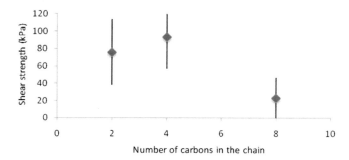

Fig. 5.14 Single lap-shear strength values of pig skin/cyanoacrylate monomer/pig skin adhesive joints. Values obtained 60 s after joint formation

by increasing the length of the alkyl hydrocarbon chain of the cyanoacrylate as a consequence of the increase in its molecular weight. A cohesive failure in the adhesive is always obtained in all adhesive joints.

Figure 5.14 shows the single lap-shear strength values of the pig skin/cyanoacrylate adhesive joints. The shear strength values were obtained 60 s after joint formation. A similar trend in joint strength as that in Fig. 5.13 is obtained and comparable adhesion is shown in the joints produced with the ethyl and *n*-butyl cyanoacrylate.

The values obtained are high and sufficient to maintain the tissues joined in wound closure. A mixed failure (cohesive in the adhesion and adhesion to pig skin) was obtained in all adhesive joints.

5.4   Conclusions

The increase in the length of the alkyl hydrocarbon chain in the cyanoacrylate monomer increased the intensity of the CH_2 and CH_3 bands and decreased the intensity of the C–O–C bands. Similarly, the intensity of the signals corresponding to the protons of the C=C bond in the cyanoacrylate monomer decreased and the signal of the CH_2 groups became less marked by increasing the length of the alkyl hydrocarbon chain. On the other hand, the signals corresponding to the C=C, C=O, and C≡N groups displaced to higher values by increasing the length of the alkyl hydrocarbon chain in the cyanoacrylate monomer. The TGA curves of the three cyanoacrylate monomers show several decomposition steps, the first decomposition was found at about 120 °C and the weight loss decreased by increasing the length of the alkyl hydrocarbon chain. Whereas the ECN was mainly composed of monomer, the BCN contained monomer and some dimmers, and the OCN had only a minor content of monomer showing five thermal decompositions.

Polymerization of the monomers modified the chemical structure of the cyanoacrylate. The polymerization produced the disappearance of the C=C bands as the polymerization of the cyanoacrylate was produced in these C=C bonds. Furthermore, after polymerization of the cyanoacrylate the intensity of the CH_3 group decreased, the intensity of the band due to the cyanoacrylate group was strongly reduced, and two intense new bands at 856 and 1,240 cm^{-1} due to C–O–C groups appeared. The thermal properties of the cyanoacrylate polymers also differed from those of the corresponding monomers. In general, the temperature at which decomposition was produced increased by increasing the length of the alkyl hydrocarbon chain of the cyanoacrylate. Whereas the ECN polymer showed a unique thermal decomposition at 199 °C, both the *n*-butyl and *n*-octyl cyanoacrylate polymers showed two thermal decompositions corresponding to two different structures. The SEM micrograph of the ECN polymer showed a homogeneous and smooth surface, whereas the *n*-butyl and, more markedly, the *n*-octyl cyanoacrylate polymers had a heterogeneous topography consisting in shredded polymer chains. This is in agreement with the loss in elastic modulus and cohesion of the cyanoacrylate by increasing the length of the alkyl hydrocarbon chain [16].

Adhesion of the cyanoacrylate adhesives was obtained from single lap-shear tests of aluminum/cyanoacrylate and pig skin/cyanoacrylate adhesive joints. The single lap-shear strength values decreased by increasing the length of the alkyl hydrocarbon chain, likely due to an increase in the molecular weight from ethyl to *n*-octyl cyanoacrylate. The values obtained in the joints produced with ethyl and *n*-butyl cyanoacrylates were high and sufficient to maintain the tissues jointed during wound closure.

Acknowledgements Financial support of the Spanish Research Funding Agency (MICYNN)—PET2008-0264 and MAT2009-10234 projects—is acknowledged. Authors thank Mr. Juan Carlos Lillo-Garrigós for pig skin supply and Mr. Javier Martínez-Aniorte for helping with adhesion tests.

References

1. Coover HW, Joyner FB, Sheare TH, Wicker TH (1959) Chemistry and performance of cyanoacrylate adhesives. Soc Plast Eng J 15:413–417
2. Ardis AE (1949) US patent nos. 2467926 and 2467927
3. Dossi M, Storti G, Moscatelli D (2010) Synthesis of poly(alkyl cyanoacrylates) as biodegradable polymers for drug delivery applications. Macromol Symp 289:124–128
4. Vinuela FV, Debrun GM, Fox AJ, Girvin JP, Peerless SJ (1982) Dominant-hemisphere arteriovenous malformations. Am J Neuroradiol 4(4):959–966
5. Singer AJ, Thode HC (2004) A review of the literature on octylcyanoacrylate tissue adhesive. Am J Surg 187(2):238–248
6. Eaglstein WH, Sullivan T (2005) Cyanoacrylates for skin closure. Dermatol Clin 23(2):193–198
7. Singer J, Quinn JV, Hollander JE (2008) The cyanoacrylate topical skin adhesives. Am J Emerg Med 26(4):490–496
8. Forssell H, Aro H, Aho AJ (1984) Experimental osteosynthesis with liquid ethyl cyanocrylate polymerized with ultrasound. Arch Orthop Trauma Surg 103(4):278–283
9. Bozkurt MK, Saydam L (2008) The use of cyanoacrylates for wound closure in head and neck surgery. Eur Arch Otorhinolaryngol 265(3):331–335
10. Vote BJT, Elder MJ (2000) Cyanoacrylate glue for corneal perforations: a description of a surgical technique and a review of the literature. Clin Experiment Ophthalmol 28(6):437–442

11. Seewald S, Sriram PVJ, Naga M, Fennerty MB, Boyers J, Oberti F, Soehendra N (2002) Cyanoacrylate glue in gastric variceal bleeding. Endoscopy 34(11):926–932

12. Pascual-Sánchez V, Mahıques-Bujanda MM, Martín-Martínez JM, Alió-Sanz JL, Mulet-Homs E (2003) Synthesis and characterization of a new acrylic adhesive mixture for use in ocular strabismus surgery. J Adhes 79:1–23

13. Cañizares Grupera ME, Carral Novo JM (2001) Empleo del alquilcianoacrilatos en suturas quirúrgicas. Rev Cubana Med Milit 30(1):21–26

14. Huang CY, Lee YD (2006) Core-shell type of nanoparticles composed of poly[(n-butyl cyanoacrylate)-co-(2-octyl cyanoacrylate)] copolymers for drug delivery application: synthesis, characterization and in vitro degradation. Int J Pharm 325(1–2):132–139

15. Charters A (2000) Wound glue: a comparative study of tissue adhesives. Accid Emerg Nurs 8:223–227

16. Senchenya NG, Guseva TI, Gololobov YuG (2007) Cyanoacrylate-based adhesives. Polymer Sci 49(3):235–239

17. Park DH, Kim SB, Ahn K-D, Kim EY, Kim YJ, Han DK (2003) In vitro degradation and cytotoxicity of alkyl 2-cyanoacrylate polymers for application to tissue adhesives. J Appl Polymer Sci 89(2):3272–3278

18. Rosado-Balmayor E, Almirall-La Serna A, Alvarez-Brito R (2002) Síntesis y caracterización de cianoacrilatos de alquilo con propiedades adhesivas. Introducción al estudio de su comportamiento térmico. *Revista Iberoamericana de Polímeros* 3–4:1–20

Chapter 6
Development of Bioabsorbable Interference Screws: How Biomaterials Composition and Clinical and Retrieval Studies Influence the Innovative Screw Design and Manufacturing Processes

Iulian Antoniac, Dan Laptoiu, Diana Popescu, Cosmin Cotrut, and Radu Parpala

6.1 Introduction

Worldwide annual reconstructive surgery for ligament injury is quoted as hundreds of thousands of interventions estimated up to 100,000 cases only of arthroscopic-assisted anterior cruciate ligament (ACL) reconstruction [1].

The system of interference fixation with screws is a widely accepted technique during knee-ligament reconstruction; it is important for the current accelerated rehabilitation protocols to secure grafts both in the femur and tibia.

Kurosaka et al. [2] described some of the concepts of interference screw fixation with a custom designed, fully threaded 9.0-mm cancellous screw with increased stiffness and linear load compared with initial Lambert's 6.5-mm cancellous AO screw [3].

First generation of the interference screws used for graft fixation were made by metallic biomaterials but they generate early and late postoperative complications like pain requiring implant removal [4], intra-articular migration [5], and difficulty in evaluating the postoperative evolution because metal screws distort postoperative imaging. These problems led to the development of bioabsorbable screws which provide similar fixation strength to that of metal interference screws. Many clinical reports compare bioabsorbable and metallic interference screws and show similar results in clinical performance like range of motion, knee effusion, and activity level [6–13].

I. Antoniac (✉) • D. Popescu • C. Cotrut • R. Parpala
University Politehnica of Bucharest, Spl. Independentei 313, sect. 6, Bucharest, Romania
e-mail: antoniac.iulian@gmail.com; dian_popescu@yahoo.com;
cosmin.cotrut@upb.ro; radu.parpala@gmail.com

D. Laptoiu
Colentina Clinical Hospital Bucharest, Street Stefan cel Mare 19-21, sector 2,
Bucharest, Romania
e-mail: danlaptoiu@yahoo.com

I. Antoniac (ed.), *Biologically Responsive Biomaterials for Tissue Engineering*,   107
Springer Series in Biomaterials Science and Engineering 1,
DOI 10.1007/978-1-4614-4328-5_6, © Springer Science+Business Media New York 2013

The current development of resorbable biomaterials provided the support for improvement of these implants. Bioresorbable screws became a golden standard, with growing use in additional sports medicine and traumatology techniques as patella stabilization, or different tenodesis including shoulder and ankle surgery.

Experimental and clinical papers have shown that direct graft fixation achieved with bioresorbable interference screws provides immediate stable fixation and promotes good healing by applying compressive forces to the cancellous bone within the bony tunnels [14–17].

Faster rehabilitation protocols, currently in use, are developed due to permission for early postoperative weight bearing. Graft fixation is considered to be the weak link in early and aggressive postoperative rehabilitation of ACL reconstruction.

What makes such screws suitable for so many applications with various requirements? One can only remind some of the required properties: the screws should provide enough strength until the surrounding tissue has healed and must not induce inflammatory, toxic, or carcinogenic response, being metabolized after performing the mechanical stability purpose, ideally leaving no debris. The surgical technique should be consistently reproducible, not leading to damage to the graft or surrounding tissues. The screw should also demonstrate good shelf life, as well as easy sterilization. The mechanical properties depend primarily on the structural aspects of the materials and processing technology.

An ideal biodegradable material for interference screw must initially provide secure fixation of the graft and assure a gradual degradation as biologic fixation is established, should have appropriate mechanical properties, would not cause inflammation or other toxic response, would be metabolized once it was destroyed completely, and should be sterilized and easily processed into a final product that has an acceptable shelf life.

On the other hand, some papers reveal possible disadvantages for the use of bioabsorbable fixation including inflammatory reactions, osteolysis with cyst formation, sterile sinus tract formation, and a time dependent deterioration of the mechanical properties. Common complications described in literature include early screw breakage during insertion or late fracture due to mechanical constraints, with eventual migration of biopolymer fragments [18, 19]. As observed, such cases are rare compared to the wide usage of implants and may be more related to the surgical technique than to the properties of the biomaterial (e.g., a low implant diameter or a smaller screw which does not provide adequate stability).

In general, most commercially available biodegradable interference screws used in clinical practice are chemically based on polylactide (or polylactic acid (PLA)) and polyglycolide (or polyglycolic acid (PGA)), but others include polycaprolactone (PCL), polydioxanone (PDO), and poly(trimethylene carbonate) (TCM). For decades, these polymers have been used for clinical applications as sutures, osteosynthesis implants (plates, screws), and drug delivery devices [20, 21].

There is also a number of biodegradable polymers derived from a natural source such as polysaccharides (cellulose, chitin, dextrin) or modified proteins that can be used as raw materials for interference screw, but synthetic polymers have some advantages over these [20].

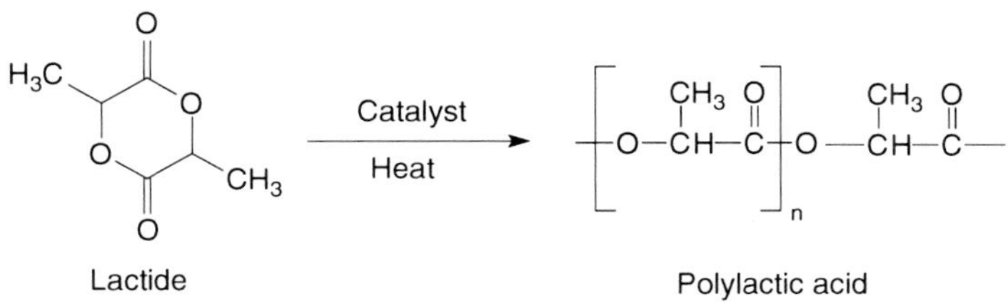

Fig. 6.1 The polymerization of lactide to polylactic acid

In general, polymer degradation is accelerated by an increased hydrophilic tendency of the polymer backbone or in the extremities, increased reactivity among hydrolytic groups in polymer chain, a low crystallinity, more pronounced porosity (larger pore sizes), and lower ends.

A polymer is generally named based on the monomer it is synthesized from. For both lactic acid and glycolic acid, an intermediate cyclic dimer is prepared and purified, prior to polymerization. These dimers are called lactide and glycolide, respectively.

PLA is an aliphatic polyester, which can be easily produced through ring-opening polymerization of lactide using a stannous octoate catalyst (Fig. 6.1).

PLA has chiral carbon in its structure and thus can exist as two stereoisomers, called poly-L-lactic acid (PLLA) and poly-D-lactic acid (PDLA), or as a racemic mixture called poly-DL-lactic acid (PDLLA). Most current interference screws are from PLLA, a polymer with semicrystalline structure, a glass transition temperature between 55 and 60°C, and a melting temperature between 173 and 178°C [22] which has good strength and long degradation period (2–5 years). Instead poly-DL-lactide has an amorphous structure with a lower strength and shorter degradation period (12–16 months) making it a very interesting material for drug delivery application. Copolymers of PLLA with PDLA or PDLLA have also been studied because when adding d-isomers the degradation rate changes as a consequence of the increase of amorphous regions and decrease of crystallinity. The properties of PLLA and its copolymers can be affected by factors related to material, such as polymerization method, L/D ratio, and thus molecular weight, polydispersity, degree of chain orientation, and by factors related to implant like porosity, size, and shape of the implant [23].

The PGA is a stronger acid, and highly crystalline (around 50%) with a glass transition temperature between 35 and 40°C and melting point of 225–230°C [24]. The ring-opening polymerization (Fig. 6.2) of the cyclic diester of glycolic acid allows the obtaining of polymers with high molecular weight, containing 1–3% residual monomer. According to the data from scientific literature it fully degrades in 3–6 months [25, 26].

PGA is more hydrophilic and thus more susceptible to hydrolysis than PLA; even the screws made by PGA show similar results regarding the clinical results, with a tendency to faster degradation. A clinical review of these models concludes that the rate of polymer degradation can be related to the size, geometry, and, finally, the design of the implant [25].

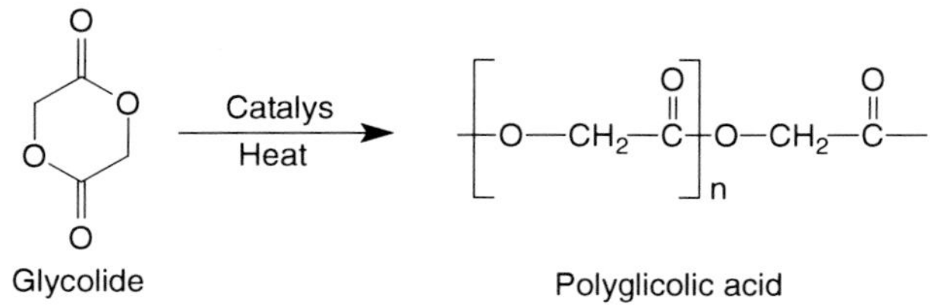

Fig. 6.2 The polymerization of glycolide to polyglycolic acid

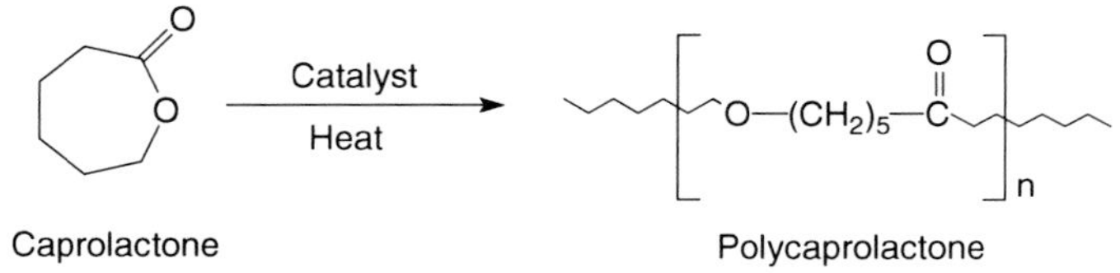

Fig. 6.3 Ring-opening polymerization of ε-caprolactone to polycaprolactone

In addition to lactic acid copolymers, a combination of PLA and PGA is an effective alternative in developing orthopedic implants. PLGA has excellent biodegradability, biocompatibility, and nontoxic properties [26]. Given the different degradation rate of PLA and PGA degradation in the case of lactic acid–glycolic acid copolymer (PLG) may be modified based on the PLA/PGA ratio. Copolymers of PLG frequently used in orthopedic devices have ratios as PLA/PGA = *50:50, 75:25*, and *85:15*.

PCL is one of the most flexible and easy-to-process biodegradable polymers, prepared by ring opening polymerization of ε-caprolactone in the presence of stannous octoate as catalyst (Fig. 6.3). PCL exhibits several unusual properties not found among the other aliphatic polyesters. These include its exceptionally low glass transition temperature (Tg) of −60°C, low melting temperature (Tm) of ~60°C, and high thermal stability with a decomposition temperature (Td) of ~350°C, while other polyesters decompose at ~250°C. PCL is semicrystalline with a low modulus and about 2 year's degradation time [27]. Due to the combination of crystallinity and high olefinic character of PCL homopolymer the hydrolysis process is considerably slower than the other poly(α-hydroxy esters) such as PGA and PLA.

Several properties of these polymers are detailed in Table 6.1.

In order to obtain biodegradable composite materials for interference screws, the same synthetic biodegradable polymers mentioned below are used as matrix reinforced with biodegradable ceramics (hydroxyapatite and tricalcium phosphate (TCP) are used generally as ceramic filler). In the case of composite materials used for interference screw, addition of the ceramic filler can have several advantages because the filler is osteoconductive and increases bone bonding to the implant and basic fillers may also buffer the acidic degradation products from the polymer, reducing the chance of foreign body reactions and increasing the mechanical properties.

Table 6.1 Properties of some biodegradable polymers used in clinical practice

Polymer abbreviation and name		Tg (°C)	Crystallinity (%)	Degradation time (month)	Elastic modulus (GPa)	Elongation (%)	Notes
PGA	Polyglycolic acid	35°C	50%	3–6 months 50% strength lost at 2 weeks 100% strength lost at 4 weeks	7.0	15–20%	– First used as biodegradable suture (DEXON®) – Strong, stiff – Rapid degradation may be limiting – Used for rods
PLLA	Poly-L-lactic acid	58°C	40%	>24 months (up to 60 months)	2.7	5–10%	– Used for interference screw, suture anchor, meniscus repair – May be self-reinforced and combined with biodegradable ceramic filler for improved strength and improved bone growth
PDLLA	Poly-D,L-lactic acid	55°C	0%	12–16 months	1.9	3–10%	– As a homopolymer, generally used for drug delivery due to low strength
LPLG	Poly-L-lactic acid-co-glycolic acid	–	0%	5–6 months	–	–	– 25–70% glycolic acid will disrupt the crystallinity of PLLA – Used for suture anchors, screws, and plates
DLPLG	Poly-D,L-lactic acid-co-glycolic acid	85/15 ~50°C 75/25 ~50°C 50/50	0%	1–5 months	~2.0	3–10%	– Copolymer of 50/50 DL-lactic acid and lactic acid – Will decompose faster than either PGA or PDLLA – Not crystalline like PGA – More hydrophilic than PDLLA – Used for suture anchors, interference screws, plates
PCL	Polycapro-lactone	~-60°C	<40%	>24 month	0.4	300–500%	– May be copolymerized with PGA to reduce PGA's stiffness – Used for sutures, pins
PDS (or PDO)	Polydioxanone	~-10°C	<40%	6–12 months	1.5	–	– Used for fracture fixation pins

Hydroxyapatite [$Ca_{10}(PO_4)_6(OH)_2$] has a definite composition, Ca/P weight ratio of 2.151, Ca/P molar ratio of 1.67, and a crystallographic structure [28]. Since hydroxyapatite is the main mineral constituent of teeth and bones, it seems to be a very suitable biomaterial for the hard tissue implants and for scaffolds used in hard tissue regeneration. Hydroxyapatite ceramics do not have any toxic effects. Shikinami and Okuno [29] describe the possibility to increase the mechanical properties of biodegradable polymers using the combination with hydroxyapatite (HA) and Yasunaga et al. [30] mention that the inclusion of HA into PLA has shown an advantage for composite materials regarding the bone bonding ability.

TCP is another biodegradable ceramic that proved to be bioactive and useful for obtaining composite biomaterials. TCP occurs in four polymorphs α, β, γ, and super α, β-TCP being preferred to the other forms due to its chemical stability, mechanical strength, and more predictable and consistent bioresorption rates [31]. The γ polymorph is a high-pressure phase and the super α polymorph is observed only at temperatures above 1,500°C. Because the degradation rate of β-TCP is higher than that of HA, the partial degradation of calcium phosphates stimulates the bone–bioceramic bonding. HA and β-TCP present differences in chemical composition and crystallographic structure. This aspect has a major impact on their physical characteristics [32].

The composite materials used for interference screws keep all advantages of bioresorbable polymers and assure new abilities like the regulation of the rate of degradation and neutralization of the degradation by-products, less failure during insertion, and improved bioactivity and osteointegration.

But one of the major problems for manufacturing of the composites is the agglomeration of the HA and β-TCP powders in the polymeric matrix. In general, fine particles tend to combine together, through electrostatic or van der Waals forces, to form agglomerated particles. This agglomeration tends to decrease the mechanical properties of the composite. So, the selected technique for preparing the composites must have the ability to break down agglomerated powder and disperse it into the polymeric matrix [33].

Another advantage of preparation of composites would be that polymers are generally weak mechanically and the addition of a ceramic component would enhance their strength and stiffness for more functional applications.

Different available composite interference screws are presented in Table 6.2. A quick market analysis revealed to us that all major companies on the market have a composite interference screw and this is more expensive than similar screws made by biodegradable polymers. Also, each company uses a proprietary composition of composite materials, which differs from other competitors.

What changes can be made in the design and manufacturing of interference screws, in order to improve graft fixation, is also under current research. Also, several surface biofunctionalization issues could be very important for the interaction phenomena between screw and tissue, with strong effects on the clinical performance of the interference screws.

In our experience, the degradation process is affected by many implant-related factors starting with biomaterial type and ending with thread design and size. Also, most of the biodegradable screws used in current practice do not promote bony regrowth.

Table 6.2 Different available composite interference screws

Composite materials		Product/manufacturer
Polymeric matrix	Ceramic phase	
70% PLDLA	30% BCP	Biocomposite/Arthrex
70% PLLA	30% β-TCP	Biolok/Arthrocare
40% PDLLA	60% β-TCP	Composit/Biomet
75% SR PDLLA	25% β-TCP	Matryx/ConMed Linvatec
70% PLLA	30% β-TCP	Bio-Intrafix/De Puy Mitek
70% PLGA	30% β-TCP	Milagro/De Puy
75% PLLA	25% HA	Biosure/Smith & Nephew
75% PLLA	25% HA	Biosteon/Stryker

6.2   Interference Screw Classification and Design

A thorough analysis of the bioresorbable interference screws currently on the market allowed a classification based on their geometrical specifications and design:

* Non-cannulated, cannulated, cannulated and perforated
* Nonself-tapping, self-tapping

Overall shape:

* Tapered (conical)
* Cylindrical
* Cylindrical + conical
* Shape of the channel for screwdriver: Hexagonal, triangular, quadrangular, six-stars, tri-lobe, turbine-like
* Number of flutes for tapping: 1, 2, 3, or 4
* Screw head: Round head, tapered head, flat head (headless), hemispherical
* Single helix thread, double-helix thread
* Thread profile: Buttress, V-shaped
* Pitch: Constant, variable

Different types of bioresorbable interference screw   hown in Table 6.3 in order to illustrate some of the above-mentioned design c   ristics and the manner in which the producers leading the market in this fiel   :hosen to adapt the design to the ACL clinical requirements.

As it can be seen, the designs of these screws do not differ   icantly, the variations being given by the thread shapes and sizes, values of the taper angle, or tip geometry. Moreover, the producers provide almost the same range of dimensions for the interference screw.

The usual range of dimensions (length and major diameter) varies between 8 and 10 mm as major diameter (for some interference screw models 7, 11, and 12 mm diameters are also available) and 25–35 mm in length (screw with lengths of 20 and 23 mm can also be purchased for special indications). Obviously, this range is

Table 6.3 Different types of bioresorbable interference screws analyzed

Cannulated/ perforated	Self-tapping/ number of flutes	Shape	Shape of channels for screwdriver	Screw head	Thread shape/ single helix	Name/producer	
Perforated	No	Cylindrical and tapered at the distal end	Cross hollow core; allows to be filled with graft	Round	Constant pitch/ single helix/ buttress thread combined rounded on the cylindrical part	BioCore— Biomet	
Cannulated	No	Conical	Hexagonal (additional triradiate for round head)	Headless (round for femoral screw)	Blunt threads	Delta Bio Interference Screw Arthrex	
Cannulated	No	Cylindrical/ cone distal	Hex	Round	Pointed with a starter thread, chamfered	Milagro— DePuy Mitek	

	Cannulated	No	Cylindrical	Star (torx)	Round	Reverse threads for right knee	Biosure—HA—Smith & Nephew
	Cannulated	No	Wedge tip	Square	Small head	Fully threaded	Biosteon—Stryker Endoscopy
	Cannulated	No	Cylindrical and tapered at the distal end	Tri-Lobe driver interface	Low-profile rounded head	Constant pitch	Matryx—Conmed Linvatec
	Cannulated	Self-tapping/1 flute	Tapered		Headless	Variable pitch/single helix/fully threaded	Biotrak—Acumed

(continued)

Table 6.3 (continued)

Cannulated/ perforated	Self-tapping/ number of flutes	Shape	Shape of channels for screwdriver	Screw head	Thread shape/ single helix	Name/producer	
Cannulated	No	Cone-shaped distal end	Hex	Headless, round head models	Constant pitch, double helix, buttress thread	MisBio—Orthomed	
Cannulated	Self-tapping/1 flute	Tapered	Disposable procedure kit	Round	Constant pitch/ single helix	IFix PEEK—Cayenne Medical	

determined by the femoral tunnel, whose length usually varies between 25 and 30 mm with an inner diameter of 8–10 mm (8 mm for quadrupled semitendinosus-gracilis graft and 10 mm for bone-tendon-bone-patellar graft).

Potential pitfalls of interference screw fixation include incorrect screw size, graft/screw tunnel mismatch, tunnel/screw divergence or convergence, graft advancement, graft translocation, graft fracture, and laceration of tensioning sutures or tendons-graft. Ideal placement of the interference screw within the femoral osseous tunnel is anterior to the graft, which allows the collagen fibers to be positioned posteriorly in the most anatomic position [34]. The optimal orientation of the interference screw within the tunnel for maximum fixation strength is parallel to the graft. If the screws diverge or converge, fixation strength may be compromised [35].

General specifications for interference screws, concerning both material and design, are dictated by the specific clinical needs and ligament reconstruction techniques. Thus, an "ideal" interference screw must minimize tissue laceration and shear stress on graft, while maximizing insertion torque pullout strength and interference with the driver. In the case of an interference screw manufactured from a bioresorbable material, the requirements for material refer also to the absorption rates and biodegradability correlated to osteointegration.

In the same time, the screwdriver (its outer shape being the negative of the interior of the screw) has to be designed to avoid the screw breakage during insertion, to minimize the insertion torque, and to distribute it along the whole screw length.

The literature includes a large number of biomechanical studies related to the screws for ACL reconstruction, focused on establishing relationships between various factors/parameters such as bone mineral density, bone block size, divergence angles between screw and bone plug, interference screw diameter, shape, thread and length, gap between tunnel and bone block, etc., and fixation strength, pullout strength, or insertion torque [13, 36–48].

Another group of biomechanical tests was used to optimize the screw design by conducting analyses which evaluate the influence of different critical design parameters (such as major/minor diameter, thread shape, or pitch) on the pullout strength, shear, tensile and compressive forces, and insertion torque. The conclusions of all these researches must be correlated to the type of material/specimens (human, sawbones, bovine, ovine, porcine) used in the experiments [44, 49].

There are also papers which use finite element simulations for analyzing different designs of interference screws or for optimizing the tunnel placement [38, 41, 50–52]. These studies assume that the materials for cortical and cancellous bones are linear elastic, isotropic, and homogenous, the mechanical properties being determined experimentally [53–56]. A more recent approach, developed with the purpose to improve the estimations made regarding the bone material properties, is to assign material properties to bones based on the relation between Hounsfield units, bone mineral density (information available by quantitative computer tomography (QCT)), and elastic modulus. QCT gives the real spatial distribution of bone density and according to these values, material properties are mapped on the finite element model [57–59].

Our observations of these studies are further presented:

6.2.1   Major and Minor Diameters of the Screw

The thread depth or the ratio between minor and major diameters determines the contact area between the screw and the bone, thus influencing bone regrowth. Asnis et al. [37] demonstrated that the root diameter has an important effect on the holding power, while Chapman et al. [40] experimentally studied different factors affecting the pullout strength, the major diameter, and the thread depth being critical in this sense. In another experimental study, DeCoster et al. [60] showed a roughly linear dependence between major diameter and pullout force. Costi et al. [61] experimentally confirmed the influence of screw diameter on the torsional strength, testing screws from different manufacturers subjected to manually applied torque. Tensile and compressive forces are proportional to the square of the minor diameter and shear forces are proportional to the cube of the minor diameter.

The insertion of the screw determines not only a compression of the graft, but also an enlargement of the tunnel [62]; therefore, ideally, after an initial conical shape for easy insertion, most of the screw diameter should remain constant, for intimate contact both with the graft and the tunnel. Interference screws of 9–10 mm in diameter and around 35 mm in length are used for tibial site fixation, while sizes for femoral screws are closer to 7 mm diameter by 28 mm length.

6.2.2   Thread

The shape of the thread is considered another critical influence factor for the pullout strength [40]. A general design rule states that, in order to avoid lacerations of the graft, the thread must not contain sharp edges. The thread profile ideally should be symmetrical, because the screw is subject to diametrically opposed stress on the side of the bone plug and on the side of the tunnel wall. Analysis of several thread designs [63] showed that the pullout strength is not significantly different for the analyzed types of threads: buttress vs. V-shape. However, Wang et al. [64] reported that a value of 30° for proximal half angle ensures an equal distribution of the pullout force on the recipient bone. Also, Hansson and Werke [50] used finite elements method to assess the effect of thread shape and size on the stress peaks for the cortical bone.

6.2.3   Screw Length

Black et al. [65] performed biomechanical experiments, which showed that the interference screw length does not influence in a significant manner the insertion torque, load to failure, displacement, or stiffness. However, Lima et al. [66] showed that the implant torque for both cylindrical and parallel miniscrews increases with the screw length. Herrera et al. [42] experimentally demonstrated that longer and wider interference screws provide better fixation in ACL graft fixation, especially in the tibial tunnel. In a comparative study, Walton [12] demonstrate that absorbable interference screws of 28 mm or 35 mm provide similar or even better fixation than metallic screws. Selby et al. [67] demonstrated that it is easier to place the insertion torque force near the articular line with long screws, making the graft less prone to failure mechanisms as the "windshield" effect.

6.2.4 Cutting Flutes

The effect of cutting flutes (as number and length) was investigated by several authors [38, 47, 63, 68]. Rubel et al. [69], comparing the pullout strength for regular and self-tapping screws, showed that the latter provide less pullout strength despite the designs and lengths of the cutting flutes, while Schatzker et al. [68] and Koranyi et al. [63] reported no significant difference between the self-tapped and pretapped screws in terms of pullout strength. A comparison between self-tapping screws with three short cutting flutes, and with two long cutting, showed that the pullout strength is superior for the former. Yerby et al. [47] proved that the insertion torques decrease if the length of the cutting flutes increases, for the same number of flutes, performing experiments of six different designs of self-tapping screw inserted in human cadaveric femurs.

6.2.5 Screw Core/Screwdriver Shape

The screwdriver shape has to be designed to ensure that the insertion torque will be uniformly distributed over a larger area, in order to avoid the screw breakage during the insertion.

Cannulated screws in comparison with non-cannulated screws improve the insertion torque and enhance the surface area for tissue regrowth, but at similar sizes, they have worst performances in terms of pullout strength due to a smaller thread depth [70]. However, Kissel et al. [43] showed that there is no significant difference on pullout strength between cannulated and non-cannulated screws with 2.5 mm diameter or less.

Weiler et al. [49] showed that turbine-like shape of the driver determines less stress concentration at the interface, increasing the torque at failure.

6.2.6 Pitch

A higher number of screw threads in contact with the bone and a higher thread depth increase the screw–bone interface, thus the axial shear area [37, 60, 71, 72]. Ricci et al. [46] investigated the effect of the pitch on the insertion torque and on pullout strength for osteoporotic cancellous bone. The experimental studies were performed on five different screws showing that a screw design optimized for pullout strength is not necessarily optimized also for the insertion torque.

6.2.7 Tapered vs. Cylindrical Screws

Biomechanical studies performed on sheep bones [44] showed that the tapered screws have higher pullout strength and insertion torque in comparison to the parallel screw.

6.3 New Design for a Bioresorbable Interference Screw

6.3.1 Clinical and Retrieval Study

In our clinical practice, the use of bioresorbable screws raised several technical problems regarding a tendency to pull out of the femoral bone or pull in from the tibial bone during the early postoperative rehabilitation procedures and a higher risk of fracture or thread deformation at the screw tip—due to hard bone and an absorption rate—sometimes incomplete—caused by an improper local blood flow in the bone.

Therefore, the majority of bioabsorbable composite screws, in order to avoid the screw breakage or damage during insertion, require notching or tapping as a supplementary step for preparing the implantation site. This procedure is inaccurate as it is performed in a cancellous bone, which causes difficulties in estimating the right diameter necessary for a stable ligament fixation. Therefore, elimination of the preparation of the screw insertion site offers a significant advantage by minimizing the risks of graft damage and decreasing the operation time.

Clinical studies were conducted on a group of 96 patients operated by same team of surgeons from Colentina Clinical Hospital, Bucharest, with at least 1 year followup, for Kenneth Jones type of arthroscopic ACL reconstruction: patellar bone— tendon—bone auto/allograft; with bioresorbable screws (Storz Megafix and Stryker Bioabsorbable Interference Screw, PLLA) (Fig. 6.4). Patients were followed during a prospective trial (clinical, X-ray) associated, when possible, with MRI at 6 and 12 months. A standardized Hoffman classification [73] was used to document and treat possible adverse tissue reaction, with special regard to extra-articular and intra-articular soft-tissue response (synovial reactions) and to osteolytic lesions. Retrieval analysis could be performed during two revision cases (rate 2% per year) in association with core biopsy when accepted by patient at a 1-year follow-up time.

Surface analysis of failed interference screws was made by scanning electron microscopy (SEM) technique that appears to be a valuable tool for this kind of analysis, using a Philips XL30 ESEM microscope in the SEM laboratory from University Politehnica of Bucharest. Stryker instrumentation uses Tibial Coring

Fig. 6.4 Intraoperative—arthroscopic image of the interference screw during insertion for anterior cruciate ligament graft fixation in the intercondylar notch

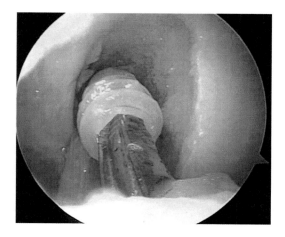

Fig. 6.5 Composite picture of a retrieved interference screw obtained from a series of SEM images

Reamers that allow us to extract the PLLA screw without damage, as seen in Fig. 6.5 which presents a complex image of the screw using a mix of different SEM images at the same scale.

SEM analysis suggested that the amount of threads engaging the bone plug differed in different screw designs. Figure 6.6 shows a progressive increase in cracks and cavitations on the root and thread with time. Cracks appear first in the thread with an axial orientation.

Also, SEM evaluation of the screw showed significant thread damage attributable to bone plug pullout (Fig. 6.7).

Another observation from the clinical practice was that the failure of the graft usually occurs at the tunnel entrance site, therefore being possible for screw threads to produce tissue laceration or lesions that cumulate during the postoperative rehabilitation.

Therefore, the SEM analysis of bioresorbable screws of PLLA showed that the screw threads present no similarity to their original shape after retrieval (even if degradation began for some time, the design should have similarities). Ultrastructural analysis showed a gradual formation of cracks and cavities, over time, starting at the top of the screw thread. It seems that the thread is due to external damage and forcing the extraction.

In general, the degradation process includes progressive increase in cracks and cavitation on the root and thread with time, which is consistent with anatomo-pathological

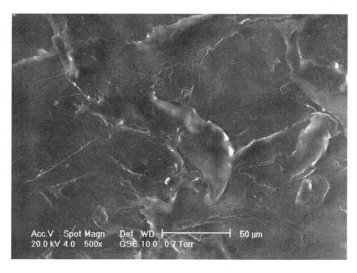

Fig. 6.6 SEM aspect of retrieved PLLA screw surface—progressive cracking of the surface

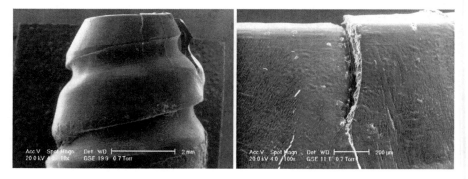

Fig. 6.7 SEM aspect of retrieved PLLA screw surface—deep fissure of the thread due to excessive torque during insertion

findings (the fragmentation of the screw allows the bone to heal inside the gaps), but the same progressive cracking of the surface could present a slight parallel orientation (slippage direction) (Fig. 6.8).

Screw degradation process occurs as a natural biodegradability delamination associated with the device within the tissue, after a while. Also, it should be noted that the screws were still present in the body in 9–12 months after implantation, even with visible signs of degradation. This is possible because the polymer will not degrade fast enough and still be influenced by mechanical degradation movements of the grafts inside the bone tunnels. MRI investigations detected no inflammatory response, but incomplete bone block incorporation was noticed at 6 months in observed patients (Fig. 6.9).

Also, all analyzed interference screws showed signs of degradation which were still clearly visible even 12 months after ACL reconstruction (Fig. 6.10).

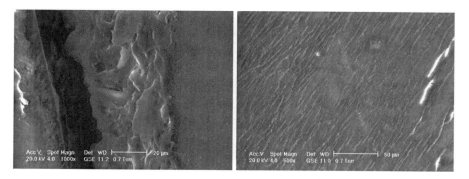

Fig. 6.8 SEM aspect of retrieved PLLA screw surface—progressive increase in cracks and cavitation on the root and thread

Fig. 6.9 MRI aspect of the bioresorbable interference screw after implantation

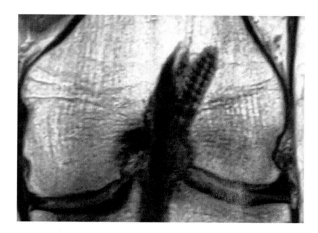

Fig. 6.10 MRI aspect of the femoral screw at 2 years with "ghosting" due to incomplete resorption of the screw

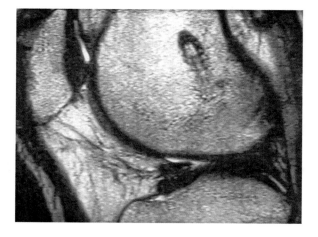

Analysis of retrieved bioresorbable interference screw confirmed that future research should be directed to the improvement of material properties using new composite materials (e.g., hydroxyapatite components which facilitate the bone in-growth). Attention should be given to the screw self-tapping design which will assure reduction of breaking forces and avoid damaging the soft tissues, especially the of the reconstruction graft.

6.3.2   Innovative Design of a Bioresorbable Interference Screw

Based on clinical experience and finite element simulations, a new bioresorbable interference screw design is proposed. The self-tapping interference screw has the following main design particularities (Fig. 6.11):

- The shape of the screw is cylindrical at the proximal end and tapered at the distal end for approximately one-third of the whole screw length.
- The thread depth of 1.5 mm (for a screw of 10 mm diameter) for the proximal cylindrical zone, progressively decreasing along the conical zone.
- The shape of the thread differs for the cylindrical and tapered zones; the passage from a shape of the thread to the other is made progressively, no sharp edges being generated; double-helix thread with constant pitch of 2 mm.
- Hexagonal interior channels with different dimensions for the cylindrical and conical zones.
- Three cutting flutes uniformly distributed on the conical part of the screw.
- Proximal rounded thread edges.

Different designs of the interference screw were obtained by varying the attack angles (the first angle of the thread flank in the insertion direction) for the distal and proximal zones (25°/30°, 25°/40°, 30°/30°, 30°/40°). In order to assess the influence of these angles on the screw holding strength, several FE simulations were performed using ANSYS v12 software (Figs. 6.12, 6.13, 6.14, 6.15, 6.16, and 6.17). The maximum displacement for each screw design and the equivalent von Misses stresses were calculated in a static structural analysis in which the tendon was loaded along the tunnel axis at 200N, according to the data presented in literature [74].

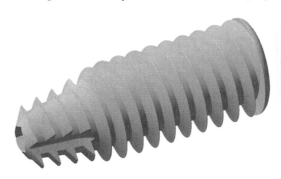

Fig. 6.11 3D CAD model of the innovative interference screw

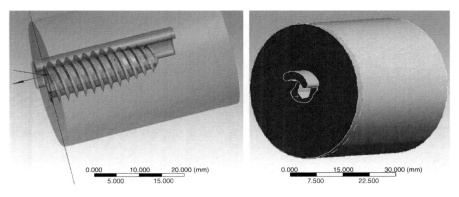

Fig. 6.12 Geometry for finite element analysis

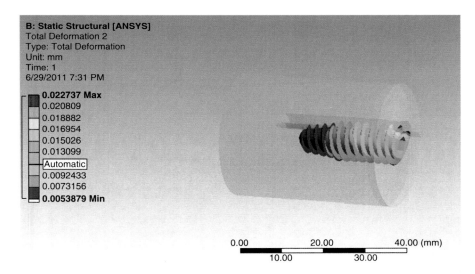

Fig. 6.13 Total deformations for 25°/30° interference screw

The bone was fixed in all three directions on the exterior surfaces, and the contacts between tendon, screw, and bone were considered bonded. For the experiment design the fact that the forces in the reconstruction graft are collinear with the femoral tunnel at 100° of flexion while the tibial tunnel remains collinear all through the range of motion was taken into account.

The mechanical properties used for screw, bone, and tendon are presented in Table 6.4. The model was meshed using tetrahedral elements with six degrees of freedom per node. The total number of nodes was 115,859, the number of solid elements was 67,105, and the number of contact elements was 24,636. The results of the finite element analysis are also presented in Table 6.5. FEM analysis showed that from the investigated angle combinations, 25°/40° provides the best holding strength.

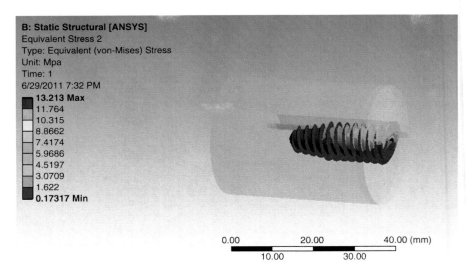

Fig. 6.14 Equivalent von-Mises stress for 25°/30° interference screw

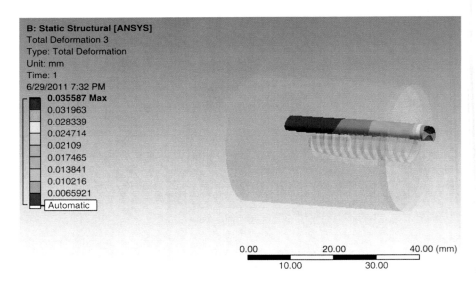

Fig. 6.15 Total deformations of the tendon for the fixation with a 25°/30° interference screw

6.3.3 Discussion About the Functionality of the New Design for Bioresorbable Interference Screw

In order to eliminate potential implant breakage (deep threads provide areas of low resistance) and maximize the press fit effect in the bone tunnel, the bioresorbable interference screw has a thread profile, which takes into account

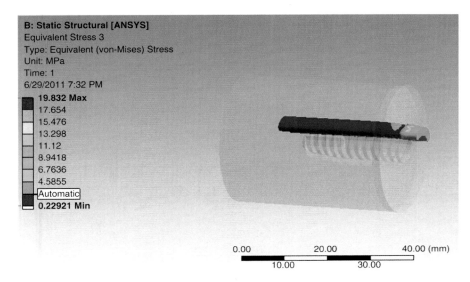

Fig. 6.16 Equivalent von-Mises stress for tendon for the fixation with a 25°/30° interference screw

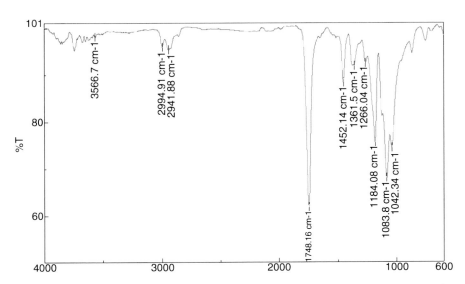

Fig. 6.17 FT-IR spectra of the biocomposite materials (70% PDLLA + 30% β-TCP) used for manufacturing innovative interference screw

the material properties, as well as the particular structure of cancellous and cortical bone.

Conical profile makes the screw easily inserted, providing also a superior fit resulting in a progressive increase in diameter. At insertion, the screw engages in the

Table 6.4 Mechanical properties for bone, interference screw, and tendon

Material	Young's Modulus (MPa)	Poisson's ratio	Density (g/cm³)
Interference screw	53,000	0.3	3
Cortical bone	13,400	0.24	2.21
Cancellous bone	283	0.29	283
Tendon	1,604	0.27	1.95

Table 6.5 FEM analysis results for interference screw

Runs	Angles distal/proximal (°)	Total deformations (mm)	Equivalent (von-Mises) stress (MPa)
1	25°/30°	0.022737	13.213
2	25°/40°	0.212561	12.112
3	30°/30°	0.022221	12.289
4	30°/40°	0.023055	16.857

cortical bone from the extremity of the bone tunnel and fills almost the whole inlet, thus providing possible initial guidance and ease of manipulation.

This self-tapping type of screw requires the distal zone of the implant to have a tapered hexagonal cannulation with a different step size so that the tip thickness, associated with the composite biomaterial properties, prevents breakage during insertion. The channel continues along the entire length inside screw, enabling the use of a nitinol guide wire for additional guidance to insert the screw into the bone. Screw is provided externally with constant pitch double thread, which is continued up to 1 mm of proximal end so that it can be fully inserted into the bone.

The proximal zone of the screw is rounded to avoid graft damaging and at the same time to reduce conflict with periarticular tissues where the integration is incomplete due to implant position.

Thread geometry is different on the two parts of the screw in order to improve contact with the bone and, consequently, resistance to tearing. Hence, the thread depth is greater for the cylinder compared to the truncated cone; it ensures better insertion in cortical bone screw. Also, the conical thread is more aggressive to penetrate more easily into cortical bone and the cylindrical part has a low thread angle in order not to damage the graft. The transition of the thread profile from one form to another is progressive, without introducing additional sharp, aggressive edges. This allows the screw to be inserted without additional tapping and results in optimum compression of the neighboring bone material.

The hexagonal tip of the screwdriver distributes torque along the entire length of the implant, so the torsional forces created do not lead to screw breakage. The favorable load transmission between screwdriver and screw permits the implant to be manufactured of the mechanically less robust composite biomaterials with polymeric matrix.

6.4   Biodegradable Interference Screw Manufacturing Processes

Weiler et al. [48] showed that torque to failure depends not only on the screw design, but also on the biomaterials composition (e.g., pure PLLA implants have a higher stiffness compared with PDLLA implants). Also, it is proved that addition of TCP modifies the mechanical properties of the implant, which becomes brittle; in order to control this aspect, the polymer must be submitted to different processing steps.

Despite the fact that details about the processes used for obtaining the bioresorbable raw materials are presented in the literature [75], one must take into account that the manufacturing processes and their corresponding process parameters influence the mechanical properties of the interference screw, which are different from those of the raw material.

However, the producers keep confidential the information about the exact composition and manufacturing processes of bioresorbable interference screw. Also, a small modification of manufacturing process parameters will affect the resorption rate. For instance, injection molding process determines a crystalline structure of the implant, which causes a slower resorption rate. On the other hand, machining allows a better tissue growth due to a rougher surface of the implant.

Even though the scientific literature reports injection molding, injection transfer molding, and compression molding as typical processes for manufacturing bioresorbable interference screws, the information referring to the process parameters (injection temperature, pressure, mold temperature, mold cooling, etc.) remains confidential for the manufacturers.

Researchers at Fraunhofer Institute for Manufacturing Engineering and Applied Materials Research in Bremen—IFAM [76]—reported in 2011 the development of a new granulate PLA and hydroxyapatite material that can be used for manufacturing interference screws using a powder injection molding process. Thus, the screws are manufactured directly using the new molding process, requiring no post-processing operations. Another proprietary manufacturing process is Micro Particle Dispersion used from DePuy Mitek for manufacturing Milagro Interference Screw using a biocomposite material (Polylactic/polyglycolic acid PLGA—reinforced with β-TCP).

Rapid Prototyping (RP) processes can be used indirectly in manufacturing interference screw by building the mold in which material is cast. To the best of our knowledge, there is no information in the literature regarding the direct manufacturing of a bioresorbable interference screw using an RP process, despite the advancement in this field, which now provides the possibility to use a larger range of processing materials. In this field, the most part of the research is now oriented in producing scaffolds with controlled porosity for tissue engineering using rapid prototyping processes [77, 78].

Based on the literature analysis and experimental research on other bioresorbable interference screws made by composite materials, we have selected as a material

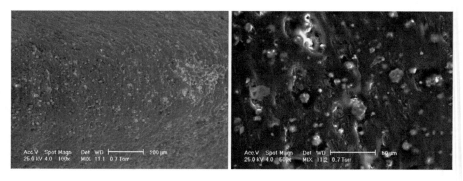

Fig. 6.18 SEM investigation of the biocomposite materials (70% PDLLA + 30% β-TCP) used for manufacturing innovative interference screw

for the innovative interference screw presented in Fig. 6.11 a composite material type PDLLA 70% and β-TCP 30%.

The obtained composite biomaterials were characterized in terms of compositional FT-IR spectroscopy on a JASCO 6200 spectrophotometer equipped with a Type A Golden Gate ATR accessory (SPECAC).

Spectra were recorded on dry samples, using a resolution of 4 cm^{-1} and an accumulation of 60 spectra. Figure 6.17 shows the FT-IR spectra characteristic of composite materials developed and confirms the presence of specific structural units of the monomer used in the organic phase (band at 1,748 cm^{-1} corresponding to stretching vibrations of the C=O group; bands at 2,941 and 2,995 cm^{-1} are attributed to the C–H group of PDLLA) and the characteristic groups of β-TCP in range of 1,100–1,000 cm^{-1} attributed to phosphate groups.

The biocomposite material structure was analyzed in detail by SEM using a Philips XL 30 ESEM microscope and worked at 25 kV and 0.7 Torr (samples not requiring coverage). The results obtained are shown in Fig. 6.18 and a relatively uniform distribution of bioceramics in polymer matrix could be observed.

Also, the EDAX profiles of the bioceramics particle (Fig. 6.19) confirm the presence of β-TCP and the convenient report Ca:P.

According to the previous considerations, for the manufacturing of the innovative interference screw we choose several manufacturing processes: machining, rapid prototyping (indirectly by building the mold as a negative of a prototype screw manufactured using a Fused Deposition Modeling (FDM) screw prototype), and microinjection molding.

Procedure FDM was developed in 1988 for obtaining prototypes in various sizes made by wax, polyester, polypropylene, and elastomer. Through this process, filaments of thermoplastic material, heated to a temperature above the melting point 1°C, are sent to a pressurized pump and then extruded through a nozzle moving in XY plane following cross-sectional shape of the model and thus filling a layer. In the next stage of the process, the platform lowers and the extrusion head make a second layer over the first. Lamellar support structures are also built along

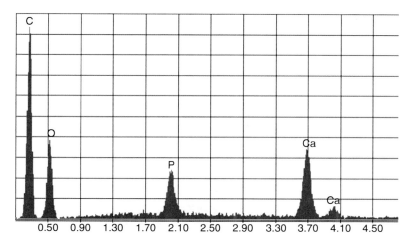

Fig. 6.19 EDAX profile of the β-TCP particle from the biocomposite materials (70% PDLLA + 30% β-TCP)

the model using a second extrusion nozzle and a different material. These water soluble structures are removed after building the prototype.

In building a prototype with FDM the process takes different steps as listed below:

- *STL file creation.* After obtaining the 3D CAD model of the part, data are converted to STL file format and read by the QuickSlice software. Process planning (part building orientation, support structure design, sectioning, path planning, and process parameter selection) is also made using the same software.
- *Sectioning.* The prototype is cut in cross sections (used values for layer thickness: 0.1778, 0.2540, 0.3556 mm) and contour curves (perimeter) are determined for each layer. Data resulting from sectioning (stored in a .SSL file) are the cross sections of the prototype.
- *Building support structures.* Depending on the geometric shape of the prototype, supports are built for sustaining the parts' structures in console and for preventing building the part directly on the platform (base support).
- *Determining FDM head path.* FDM heads move in XY plane following the planned paths. Extruded material paths have certain features, called attributes, defined by the user for each particular group of curves. After defining the necessary attributes of each set (curves having the same attributes), data is stored in a file, .SML (Stratasys Machine Language), which controls the drive wheels and motors of the FDM head and platform.
- *Prototype construction.* SML data are transmitted to the Modeler (FDM machine) and the physical prototype is built.
- *Elimination of support structure and finishing the prototype.* A cleaning system is used for eliminating the water soluble support structures after the building process is finished.

Fig. 6.20 Physical model of
the innovative interference
screw made by FDM
processes

Fig. 6.21 Silicon mold

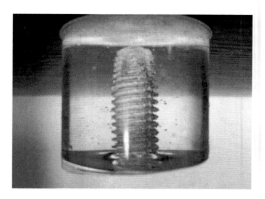

Thus, based on 3D CAD model of the screw and using a layer manufacturing technique type FDM we obtained the physical prototype presented in Fig. 6.20.

This prototype was used for making a silicone mold (Fig. 6.21). The mechanical properties of the obtained molded screw, as well as its geometric accuracy, proved unsatisfactory for the considered application because fine details could not be obtained.

The method of manufacturing which ensured the necessary mechanical properties of the screw was injection molding. Composite biomaterial obtained as a cylinder is grinded in a mill to obtain granules to be used in the injection machine. This last manufacturing process provides both precision and productivity.

6.5 Conclusions

Clinical studies of implant failures showed the importance of the screw thread and wedge designs on its engagement in the bone plug, as well as the importance of the type of bioresorbable material on the soft-tissue reactions and risk of cracks or breakage.

Biocomposite interference screws are commonly used now in the knee surgery. Initial fixation strength and ease of insertion are vital to the stability of the reconstruction of ACL, especially with soft tissue grafts and in the tibial tunnel.

In our study several new design features of an innovative interference screw were proposed in order to ameliorate press-fit fixation without damaging the graft. We proposed a self-tapping screw with conical shape and three cutting flutes at the distal end and cylindrical shape at the proximal end. The thread is different for these two zones, more aggressive at the conical zone to easily penetrate the cortical bone, while on the cylindrical zone the attack angle of the thread is bigger in order not to damage the graft. Screw design optimization was made by changing the values of the attack angles of the screw thread shapes and by determining different values of the pullout force using finite element analysis.

However, as already mentioned, it is the combination between the design and material which guarantees the integrity of the screw during insertion, the tissue regrowth, and the stability of fixation.

References

1. Bach BR Jr, Jones GT, Sweet FA, Hager CA (1994) Arthroscopy-assisted anterior cruciate ligament reconstruction using patellar tendon substitution. Two to four-year follow-up results. Am J Sports Med 22:758–767
2. Kurosaka M, Yoshiya S, Andrish JT (1987) A biomechanical comparison of different surgical techniques of graft fixation in anterior cruciate ligament reconstruction. Am J Sports Med 15:225–229
3. Lambert KL (1983) Vascularized patellar tendon graft with rigid internal fixation for anterior cruciate ligament insufficiency. Clin Orthop 172:85–89
4. Kurzweil PR, Frogameni AD, Jackson DW (1995) Tibial interference screw removal following anterior cruciate ligament reconstruction. Arthroscopy 11:289–291
5. Sidhu DS, Wroble RR (1997) Intraarticular migration of a femoral interference fit screw. A complication of anterior cruciate ligament reconstruction. Am J Sports Med 25:268–271
6. Caborn DNM, Coen M, Neef R, Hamilton D, Nyland J, Johnson DL (1998) Quadrupled semitendinosus-gracilis autograft fixation in the femoral tunnel: a comparison between a metal and a bioabsorbable interference screw. Arthroscopy 14:241–245
7. Caborn DNM, Urban WP Jr, Johnson DL, Nyland J, Pienkowski D (1997) Biomechanical comparison between BioScrew and titanium alloy interference screws for bone-patellar tendon-bone graft fixation in anterior cruciate ligament reconstruction. Arthroscopy 13:229–232
8. Johnson LL, van Dyk GE (1996) Metal and biodegradable interference screws: comparison of failure strength. Arthroscopy 12:452–456
9. Kaeding C, Farr J, Kavanaugh T, Pedroza A (2005) A prospective randomized comparison of bioabsorbable and titanium anterior cruciate ligament interference screws. Arthroscopy 21:147–151
10. Marti C, Imhoff AB, Bahrs C, Romero J (1997) Metallic versus bioabsorbable interference screws for fixation of bone-patellar tendon-bone autograft in arthroscopic anterior cruciate ligament reconstruction. A preliminary report. Knee Surg Sports Traumatol Arthrosc 5:217–221
11. Pena F, Grøntvedt T, Brown GA, Aune AK, Engebretsen L (1996) Comparison of failure strength between metallic and absorbable interference screws. Influence of insertion torque, tunnel-bone block gap, bone mineral density, and interference. Am J Sports Med 24:329–334
12. Walton M (1999) Absorbable and metal interference screws: comparison of graft security during healing. Arthroscopy 15:818–826

13. Weiler A, Windhagen HJ, Raschke MJ, Laumeyer A, Hoffmann RF (1998) Biodegradable interference screw fixation exhibits pull-out force and stiffness similar to titanium screws. Am J Sports Med 26:119–126

14. Bush-Joseph CA, Bach BR Jr (1998) Migration of femoral interference screw after anterior cruciate ligament reconstruction. Am J Knee Surg 11:32–34

15. Fabbriciani C, Mulas PD, Ziranu F, Deriu L, Zarelli D, Milano G (2005) Mechanical analysis of fixation methods for anterior cruciate ligament reconstruction with hamstring tendon graft. An experimental study in sheep knees. Knee 12:135–138

16. Fu FH, Bennett C, Ma CB (2000) Current trends in anterior cruciate ligament reconstruction. Part II: Operative procedures and clinical correlations. Am J Sports Med 28:124–130

17. Weimann A, Rodieck M, Zantop T, Hassenpflug J, Petersen W (2005) Primary stability of hamstring graft fixation with biodegradable suspension versus interference screws. Arthroscopy 21:266–274

18. Appelt A, Baier M (2007) Recurrent locking of knee joint caused by intraarticular migration of bioabsorbable tibial interference screw after arthroscopic ACL reconstruction. Knee Surg Sports Traumatol Arthrosc 15:378–380

19. Lembeck B, Wülker N (2005) Severe cartilage damage by broken poly-l-lactic acid (PLLA) interference screw after ACL reconstruction. Knee Surg Sports Traumatol Arthrosc 13: 283–286

20. Middleton JC, Tipton AJ (2000) Synthetic biodegradable polymers as orthopaedic devices. Biomaterials 21:2335–2346

21. Nair LS, Laurencin CT (2007) Biodegradable polymers as biomaterials. Prog Polym Sci 32:762–798

22. Park A, Cima LG (1996) In vitro cell response to differences in poly-L-lactide crystallinity. J Biomed Res 31:117–130

23. Rezwan K, Chen QZ, Blaker JJ, Boccaccini AR (2006) Biodegradable and bioactive porous polymer/inorganic composite scaffolds for bone tissue engineering. Biomaterials 27:3413–3431

24. Chujo K, Kobayashi H, Suzuki J, Tokuhara S (1967) Physical and chemical characteristics polyglycolide. Die Makromolekulare Chemie 100:267–270

25. Konan S, Haddad FS (2009) A clinical review of bioabsorbable interference screws and their adverse effects in anterior cruciate ligament reconstruction surgery. Knee 16:6–13

26. Vert M, Li SM, Spenlehauer G, Guerin P (1992) Bioresorbability and biocompatibility of aliphatic polyesters. J Mater Sci Mater Med 3:432–446

27. Sinha VR, Bansal K, Kaushik R, Kumria R, Trehan A (2004) Poly-ε-caprolactone microspheres and nanospheres: an overview. Int J Pharm 278:1–23

28. Hench LL, Wilson J (1993) An introduction to bioceramics, 1st edn. World Scientific Publishing, Singapore

29. Shikinami Y, Okuno M (1999) Bioresorbable devices made of forged composites of hydroxyapatite and poly L-lactde (PLLA): Part I. Basic characteristics. Biomaterials 20:859–877

30. Yasunaga T, Matsusue Y, Furukawa T, Shikinami Y, Okuno M, Nakamura T (1999) Bonding behaviour of ultrahigh strength unsintered hydroxyapatite particles/poly(L-lactide) composites to surface of tibial cortex in rabbits. J Biomed Mater Res 47:412–419

31. Famery R, Richard N, Boch P (1994) Preparation of alpha-tricalcium and beta-tricalcium phosphate ceramics, with and without magnesium addition. Ceram Int 20:327–336

32. Jarcho M (1981) Calcium phosphate ceramics as hard tissue prosthetics. Clin Orthop Relat Res 157:259–278

33. Mathieu LM, Bourban PE, Månson JA (2006) Processing of homogeneous ceramic/polymer blends for bioresorbable composites. Compos Sci Technol 66:1606–1614

34. Blum MF, Garth WP, Lemons JE (1995) The effect of graft rotation on attachment site separation distances in ACL reconstruction. Am J Sports Med 23:282–287

35. Brodie JT, Torpey BM, Donald GD 3rd, Bade HA 3rd (1996) Femoral interference screw placement through the tibial tunnel: a radiographic evaluation of interference screw divergence angles after endoscopic anterior cruciate ligament reconstruction. Arthroscopy 12:435–440

36. Abshire BB, McLain RF, Valdevit A, Kambic HE (2001) Characteristics of pullout failure in conical and cylindrical pedicle screws after full insertion and backout. Spine J 1:408–414
37. Asnis SE, Ernberg JJ, Bostrom MP, Wright TM, Harrington RM, Tencer A, Peterson M (1996) Cancellous bone screw thread design and holding power. J Orthop Trauma 10:462–469
38. Battula S, Schoenfeld A, Vrabec G, Njus GO (2006) Experimental evaluation of the holding power/stiffness of the self-tapping bone screws in normal and osteoporotic bone material. Clin Biomech 2:533–537
39. Brown GA, McCarthy T, Bourgeault CA, Callahan DJ (2000) Mechanical performance of standard and cannulated 4.0-mm cancellous bone screws. J Orthop Res 18:307–312
40. Chapman JR, Harrington RM, Lee KM, Anderson PA, Tencer AF, Kowalski D (1996) Factors affecting the pullout strength of cancellous bone screws. J Biomech Eng 118:391–398
41. Chizari M, Wang B, Snow M (2007) Experimental and numerical analysis of screw fixation in anterior cruciate ligament reconstruction. Proceedings of the World Congress on Engineering, London, UK
42. Herrera A, Martínez F, Iglesias D, Cegoñino J, Ibarz E, Gracia L (2010) Fixation strength of biocomposite wedge interference screw in ACL reconstruction: effect of screw length and tunnel/screw ratio. A controlled laboratory study. BMC Musculoskelet Disord 11:139–146
43. Kissel CG, Friedersdorf SC, Foltz DS, Snoeyink T (2003) Comparison of pullout strength of small-diameter cannulated and solid-core screws. J Foot Ankle Surg 42:334–338
44. Mann CJ, Costi JJ, Stanley RM, Dobson PJ (2005) The effect of screw taper on interference fit during load to failure at the soft tissue/bone interface. Knee 12:370–376
45. Patel PS, Shepherd DE, Hukins DW (2010) The effect of screw insertion angle and thread type on the pullout strength of bone screws in normal and osteoporotic cancellous bone models. Med Eng Phys 32:822–828
46. Ricci WM, Tornetta P 3rd, Petteys T, Gerlach D, Cartner J, Walker Z, Russell TA (2010) A comparison of screw insertion torque and pullout strength. J Orthop Trauma 24:374–378
47. Yerby S, Scott CC, Evans NJ, Messing KL, Carter DR (2001) Effect of cutting flute design on cortical bone screw insertion torque and pullout strength. J Orthop Trauma 15:216–221
48. Weiler A, Hoffman RFG, Siepe CJ, Kolbeck SF, Südkamp NP (2000) The influence of screw geometry on hamstring tendon interference fit fixation. Am J Sports Med 28:356–359
49. Weiler A, Hoffmann RF, Stähelin AC, Bail HJ, Siepe CJ, Südkamp NP (1998) Hamstring tendon fixation using interference screws: a biomechanical study in calf tibial bone. Arthroscopy 14:29–37
50. Hansson S, Werke M (2003) The implant thread as a retention element in cortical bone: the effect of thread size and thread profile: a finite element study. J Biomech 36:1247–1258
51. Hou SM, Hsu CC, Wang JL, Chao CK, Lin J (2004) Mechanical tests and finite element models for bone holding power of tibial locking screws. Clin Biomech 19:738–745
52. Hsu CC, Chao CK, Wang JL, Hou SM, Tsai YT, Lin J (2006) Multiobjective optimization of tibial locking screw design using a genetic algorithm: evaluation of mechanical performance. J Orthop Res 24:908–916
53. Ashman RB, Rho JY, Turner CH (1989) Anatomical variation of orthotropic elastic moduli of the proximal human tibia. J Biomech 22:895–900
54. Choi K, Kuhn JL, Ciarelli MJ, Goldstein SA (1990) The elastic moduli of human suchondral, trabecular, and cortical bone tissue and the size-dependency of cortical bone modulus. J Biomech 23:1103–1113
55. Standard handbook of biomedical engineering and design (2003) In: Kutz M (ed) Bone mechanics, Chapter 8. McGraw-Hill, New York, pp 1–23
56. Williams JL, Lewis JL (1982) Properties and an anisotropic model of cancellous bone from the proximal tibial epiphysis. J Biomech Eng 104:50–56
57. Bessho M, Ohnishi I, Matsuyama J, Imai K, Nakamura K (2007) Prediction of strength and strain of the proximal femur by a CT based finite element method. J Biomech 40:1745–1753
58. Jovanović JD, Jovanović ML (2010) Finite element modeling of the vertebra with geometry and material properties retrieve from CT-scan data. Facta Universitatis: Mech Eng 8:19–26

59. Perez MA, Fornell P, Garcia-Aznar JM, Doblaret M (2007) Validation of bone remodelling models applied to different bone types using Mimics. Available online at: www.materialise. com/download/

60. DeCoster TA, Heetderks DB, Downey DJ, Ferries JS, Jones W (1990) Optimizing bone screw pullout force. J Orthop Trauma 4:169–174

61. Costi JJ, Kelly AJ, Hearn TC, Martin DK (2001) Comparison of torsional strengths of bioabsorbable screws for anterior cruciate ligament reconstruction. Am J Sports Med 29:575–580

62. Buelow JU, Siebold R, Ellermann A (2002) A prospective evaluation of tunnel enlargement in anterior cruciate ligament reconstruction with hamstrings: extracortical versus anatomical fixation. Knee Surg Sports Traumatol Arthrosc 10:80–85

63. Koranyi E, Bowman E, Knecht CD, Jansen M (1970) Holding power of orthopaedic screws in bone. Clin Orthop 72:283–286

64. Wang Y, Mori R, Ozoe N, Nakai T, Uchio Y (2009) Proximal half angle of the screw thread is a critical design variable affecting the pull-out strength of cancellous bone screws. Clin Biomech 24(9):781–785

65. Black KP, Saunders MM, Stube KC, Moulton MJ, Jacobs CR (2000) Effects of interference fit screw length on tibial tunnel fixation for anterior cruciate ligament reconstruction. Am J Sports Med 28:846–849

66. Lima SA, Cha JY, Hwang CJ (2008) Insertion torque of orthodontic miniscrews according to changes in shape, diameter and length. Angle Orthod 78:234–240

67. Selby JB, Johnson DL, Hester P, Caborn DN (2001) Effect of screw length on bioabsorbable interference screw fixation in a tibial bone tunnel. Am J Sports Med 29:614–619

68. Schatzker J, Sanderson R, Murnaghan JP (1975) The holding power of orthopaedic screws in vivo. Clin Orthop 108:115–122

69. Rubel I, Fornari E, Miller B, Hayes W (2006) Are self tapping screw similar? A biomechanical study. J Bone Joint Surg Br 88-B:30

70. Bucholz RW, Jones A (1991) Fractures of the shaft of the femur. J Bone Joint Surg Am 73:1561

71. Gausepohl T, Möhring R, Pennig D, Koebke J (2001) Fine thread versus coarse thread. A comparison of the maximum holding power. Injury 32:1–7

72. Lavi A (2010) Internal fixation using cannulated screws. Technical paper available online: http://www.vilex.com/html/products/technical_training/internal_fixation_paper.html

73. Hoffmann R, Weiler A, Helling HJ, Ktek C, Rehm KE (1997) Local foreign-body reactions to biodegradable implants. A classification. Unfallchirg 100:658–666

74. Harrington IJ (1976) A bioengineering analysis of force actions at the knee in normal and pathological gait. Biomed Eng 11:167–172

75. Törmälä P (1992) Biodegradable self-reinforced composite materials: manufacturing structure and mechanical properties. Clin Mater 10:29–34

76. http://www.qmed.com/mpmn/medtechpulse/fraunhofer-implant-material-promotes-bone-growth

77. Zhou W (2010) Selective laser sintering of poly(L-lactide)/carbonated hydroxyapatite porous scaffolds for bone tissue engineering, Tissue Engineering, Chapter 9. University of Hong Kong, Hong Kong, pp 179–204. ISBN 978-953-307-079-7

78. Williams JM, Adewunmi A, Schek RM, Flanagann CL, Krebsbach PH, Feinberg SE, Hollister SJ, Das S (2005) Bone tissue engineering using polycaprolactone scaffolds fabricated via selective laser sintering. Biomaterials 26(23):4817–4827

Chapter 7
Modeling and Numerical Analysis of a Cervical Spine Unit

Mirela Toth-Tascau and Dan Ioan Stoia

7.1 Introduction

Pathological or accidental disorders of the human spine are responsible for worsening the living conditions of over 5 % of world population, and for some of them, the only way to reduce pain is surgical implantation. Surgery can heal or improve initial findings and may also alleviate, or even completely eliminate, the existing pain and dysfunctions. The continuing trends of life quality improvement in modern society play a key role in spinal implant development. In patients with spinal implants, the performances and sustainability of mechanical systems help in reducing the pain.

Decade of the Spine was initiated to promote awareness of the spine, spine care, and spine research. Over the course of the decade, this major spine initiative met its mission "to improve the quality of spinal care worldwide" [4].

The movement possibilities and the functionality of the spine are provided by its structural complexity. The spinal column (or vertebral column) extends from the skull to the pelvis and it counts 33 individual bones termed vertebrae. The vertebrae are stacked on top of each other and grouped into four regions: cervical, thoracic, lumbar, and sacral. In addition to the 33 vertebrae, the spine involves 24 intervertebral discs, 344 joint facets, 356 ligaments, 730 direct action muscles, and 730 points of insertion [41]. The 31 pairs of spinal nerves and the vascular system have to be added as well. Such a complex mobile system leads to a very complex mechanical behavior.

Bone is a living tissue exquisitely adapted to resist stress with suitable resilience, which supports the body and provides leverage for movements. At the macroscopic level the bone has two structure components: compact or cortical bone and cancellous, trabecular, or spongy bone [47]. Compact bone can be considered as a solid,

M. Toth-Tascau (✉) • D.I. Stoia
Politehnica University of Timisoara, Bd. Mihai Viteazu, No. 1, Timisoara 300222, Romania
e-mail: mirela@cmpicsu.upt.ro; ionut@cmpicsu.upt.ro

I. Antoniac (ed.), *Biologically Responsive Biomaterials for Tissue Engineering*,
Springer Series in Biomaterials Science and Engineering 1,
DOI 10.1007/978-1-4614-4328-5_7, © Springer Science+Business Media New York 2013

and cancellous bone, on the contrary, is a porous—spongy—material. Bone material is inhomogeneous because it is composed of a crystallized mineral and a fibrous organic component. When dealing with bone material, additional difficulties arise. Due to changes in bone density and architecture, the mechanical properties of bone change from one volume to the next. In addition, bone is non-linearly elastic. Thus, the bone material can be considered like a two-phase hierarchical composite structure. In order to study the biomechanical behavior of a bone, at least two different materials have to be attributed: cancellous bone for the core and cortical bone for the shell.

The structure of the vertebrae varies from one segment of the spinal column to the next, reflecting specific static and functional requirements. The cervical bones are designed to allow flexion, extension, bending, and axial rotation of the head. They are smaller than the other vertebrae, allowing a greater amount of movement.

The spinal column performs a variety of mechanical functions like head, upper limbs, and thorax sustaining; absorption, dampening, and transferring of the dynamic loads; and complex movements in physiological ranges. The smallest functional element of the spinal column is also known as a mobile segment or spine functional unit. A spine functional unit consists of two neighboring vertebrae, the intervertebral disc between them, the facet joints, and the ligaments [26].

Generally, the experimental testing of a biomechanical structure, including the spinal column or some of its segments, implies many issues. Thus, a computer simulation approach is far more reliable. In order to perform a computer simulation, a realistic 3D virtual model of the spine has to be developed. Because of the different nature of the spinal elements, each one has to be individually created. The assemblage of those elements materializes in the desirable model which must be highly realistic detailed life-size. The model should also be designed to answer specifically the studied aspects. Its predictions are valid only within the boundaries of assumptions and limitations that it incorporates [40].

A comprehensive classification of biomechanical models is presented by M. M. Panjabi in *Cervical spine models for biomechanical research* [40]. "Biomechanical models are widely used for the understanding of the basic normal function and dysfunction of the cervical spine and for testing implants and devices." Biomechanical models can be broadly categorized into four groups [40], each of them having its own recommendations and limitations:

- Physical models looking as natural specimens, made of nonorganic materials or obtained by plastination, are often used as valuable tools for patient and student education [27, 46, 53].
- In vitro models consisting of a cadaveric spine specimen are useful in providing basic understanding of spine behavior [40].
- In vivo animal models allow the modeling of living phenomena, such as fusion, development of disc degeneration, instability, etc.
- "Computer models are developed from mathematical equations that incorporate geometry and physical characteristics of the human spine and may be advantageously

used for problems that are difficult to model by other means" [40, 51, 59]. Advanced computer models are based on geometric modeling using CAD techniques and 3D image-based reconstruction using medical imaging techniques.

"Geometric modeling is the quantification of all the geometric data of an object as well as additional information on material properties in a single model" [3, 9–11, 14, 28, 44, 58].

Computer models have been traditionally used to assist in engineering for design, modeling, simulation, analysis, and manufacturing. Advanced researches in computer-aided technology, Information Technology, and Biomedicine developed many new and important biomedical applications. New CAD facilities combined with biology, engineering, and information science have evolved a new field of Computeraided tissue engineering. This emerging field encompasses computer-aided design, image processing, and manufacturing, enabling modeling of anatomical tissue, 3D anatomy visualization, identification and 3D reconstruction, tissue classification, tissue implantation, prototyping, and assisted surgical planning [48, 49, 54–56].

Anatomical modeling can be performed by classical design-parametric features, using any CAD software, such as Pro/Engineer, I-DEAS, CATIA, SolidWorks, and SolidEdge. Another possibility is to use reverse engineering to convert the real system into a 3D virtual solid. In a large perspective, *Reverse Engineering* (*RE*) refers to a process of discovering the technological principles of a device, object, or system through analysis of its structure, function, and operation. In geometric modeling, RE means to create a CAD model from an existing physical object in order to use it as a geometric model in simulations or for its manufacturing [64]. Improved anatomical models are usually generated by 3D image-based reconstruction, through high-resolution noninvasive imaging techniques, such as CT or MRI technology and medical imaging process and the 3D reconstruction techniques. There are commercially available 3D reconstruction programs such as Surgi-CAD by Integraph ISS, USA; Med-Link, by Dynamic Computer Resources, USA; and Mimics and MedCAD, by Materialize, Belgium, and surface generation software like Geomagic Studio [12, 21]. The solid generation is usually accomplished using SolidWorks, SolidEdge, or other CAD software.

The solid reconstruction is a branch of geometric modeling which refers to the geometric description of a real solid object [3]. The obtained solid models must satisfy three basic requirements:

- Be complete: The graphical model must not be an ambiguous representation.
- Have integrity: Operation on geometric models must maintain the connection of edges and points when moving.
- Provide accuracy in modeling of complex shapes.

Using of 3D anatomical models (static, kinematic, or dynamic) reduces the number of required spinal radiographs, improves the visualization, and allows measurements for better treatment planning. The 3D model is parameterized and used as a reference model for creating patient-specific anatomical models (customized models) based on dimensions extracted from radiographic images [7, 16, 31].

In the last few years some commercial programs were developed as solutions to modeling, simulation, and analysis of human body or human body parts. For example, the AnyBody Modeling System™, by AnyBody Technology A/S Denmark, is a software solution for simulating the biomechanics of a living structure working in concert with its environment which is defined in terms of external forces and boundary conditions. AnyBody performs a simulation and calculates the mechanical properties for the body-environment system. AnyBody can also scale the models to fit to any population from anthropometric data or to any individual (http://www. anybodytech.com/fileadmin/user_upload/anybody_lowres__2_.pdf). The biomechanical behavior of advanced anatomical models can be determined by numerical analysis. The Finite Element Analysis (FEA) of the anatomical systems, normal or implanted, is a current concern in medical engineering field. Most works are focused in a global evaluation of the anatomical systems under the loads [18, 61].

A comprehensive state-of-the-art and critical review of the Finite Element Applications in human cervical spine modeling is presented by N. Yognandan et al. in *Finite Element Applications in Human Cervical Spine Modeling* [67]. The authors are focused on the developments in model construction, constitutive law, identification, loading and boundary conditions, and model validation.

The Finite Element Method (FEM) is successfully applied to simulations of biomechanical systems. FEM is a method which allows advanced computations taking into account the essential features of the structure such as complex geometry, material inhomogeneity, and anisotropic mechanical properties of the bone and adjacent tissues. Using FEM, the stress and strain fields in the tissue may be estimated to predict the biomechanical behavior of the spinal segment. The scientific literature on the assessment of the bone physical parameters appears to be extensive; however, it is difficult to use basic information on physical parameters, because the mechanical data referring to the tissues are sometimes grossly inconsistent [37]. The heterogeneity of the tissue structure results from the people's age and gender, anthropometric data, and normal or pathological state, varying between individuals. Thus, the full potential of FEM is still not explored due to the absence of precise, high-resolution medical data [8]. However, the Finite Element Modeling used in the last 20 years became a standard tool for biomedical applications [25].

The FEM is generally used to understand and predict the biomechanical behaviors of vertebrae (taking into account trabecular bone and cortical bone of the vertebrae), intervertebral discs, ligaments, facet joints, and muscles of the spinal segments, to analyze quasi-static and dynamic applications. The FEM also includes complicated porous tissue structures, nucleus pulpous, annulus fibrosus, nonlinear ligaments, and non-thickness contact elements [36, 60, 65]. In FEA the tissue's biomechanical properties are adopted from literature or from experimental studies. When the biomechanical structure studied by numerical analyses is a certain spinal unit implanted or not, the internal balance of various components should be determined [66].

Validation of the Finite Element model is an important and challenging process. The results given by the Finite Element model must be compared with the results of different experimental studies. Thus, new experimental studies suitable for comparisons with the model are necessary in order to ensure that the model behavior is realistic [25].

Combining in vitro tests or mathematical modeling with FEA provides a more complete picture of the biomechanical behavior, both of tissues and implants.

Finite Element models developed for a certain patient could be clinically useful in the assessment of spine fracture risk, involving parametric investigation of the factors that contribute to fracture risk. The effect of changes in material properties or geometry of the intervertebral disc determined by its degeneration could be quantified using numerical analysis [15].

The main objective of this study was to obtain a sufficiently accurate model of a functional unit of human cervical spine, both implanted and non-implanted, and to analyze the biomechanical behavior of the implanted model, using FEM.

Based on CT scans and 3D reconstruction techniques a multi-solid model that accurately reproduces the geometry of the vertebra was developed. In order to achieve this objective, a cadaveric third cervical vertebra was scanned using Siemens SOMATOM Plus 4 Power system in Medical Imaging Laboratory of CMPICSU Research Centre in Politehnica University of Timisoara. The 2D DICOM images were imported into *Mimics* software and processed using Geomagic Studio software to obtain surface patches and SolidWorks CAD software to accomplish the solid generation phase. In this way, the vertebral body was created taking into account the material inhomogeneity and underlying the importance of the density influence. The main idea was to create two different solid bodies: one for the spongy core of the vertebra and another one for the cortical shell. By combining the two solids, a multi-solid body can be obtained. The advantage of using multi-solid bodies is that for each constitutive element different material properties can be set.

Using the reconstruction steps, the obtained solids satisfy all of the basic requirements, and they can be assembled into a multi-solid body. The proposed model has the following advantages:

- Provides accurate information about the vertebra size, both for bone structures and soft tissues
- Allows the determination of sections in different planes
- Allows determining of joint centers and axes of motion
- May impose motion constraints
- Allows determining of motion laws of joints
- Allows allocation of inertial characteristics (mass, moments of inertia) and material characteristics of layers
- Allows stress and strain evaluation for different components, based on numerical analysis

The evaluation of stress and strain distribution occurring in a real solid object can prevent the unexpected changes in mechanical behavior or even failure. Because an experimental analysis of the human spine behavior involves many issues, a virtual simulation of the mechanical behavior is more accessible. For this reason, the solid models of the anatomical spine segment and the mechanical stabilization system were created.

Using CAD techniques (*feature base design*), a second model of a functional unit of human cervical spine, both implanted and non-implanted, was developed. The developed model guarantees that the functional unit geometry can be imported and easily handled within ANSYS software.

The second model of the cervical unit is anatomically detailed and partially differentiates between the cervical tissues. Structures implemented in the FE model are the cancellous and cortical bone of the vertebra, intervertebral disc, cervical ligaments, and facet joints. The neck muscles were not considered.

The mechanical elements used in implantation of the functional unit consist in a stabilization system, which has the role of rigidly fixing two vertebrae (C2–C3). This stabilization system is composed of the following elements: one bone plate, four bone screws, and four locking screws [22].

In order to complete the task, a good fundament concerning the behavior of the materials under the stress condition was made. The FEM was used to determine the stresses and strains acting at the bone–screws interfaces, during a flexion movement of the head. In this way, the implanted cervical unit (ICU) was analyzed in three cases. For the three units, the bone screw type is responsible for the differences.

7.2   Solid Reconstruction Modeling of a Cervical Vertebra

A functional spinal unit or a spinal motion segment is the smallest physiological motion unit of the spine to exhibit biomechanical characteristics similar to those of the entire spine [63]. The cervical functional unit is a structure (virtual or real) composed of two adjacent vertebrae, the intervertebral disc and all adjoining ligaments.

The objective of the solid reconstruction modeling was to create a multi-solid model of a cervical functional unit. The central idea was to build a model of cervical vertebra as an assembly of two different solid bodies: one for the spongy core of the vertebra and another one for the cortical shell [50]. In this way, each constitutive element has its own material features. The inhomogeneous material of the vertebra is modeled as two homogeneous materials describing the two main components of the bone tissue: the hard outer layer of bone composed of compact bone tissue and the inner spongy structure that resembles honeycomb composed of cancellous bone tissue.

7.2.1   Image Acquisition

In order to acquire the slice images of a cadaveric cervical vertebra, a Computer Tomography (CT) system was used. The CT scanning is a painless imagistic procedure that allows visualization of the living anatomical structures by exposing the

Fig. 7.1 Overlapped DICOM
images in three views

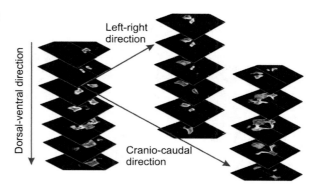

body to a controlled X-ray emission. A cadaveric third cervical vertebra was scanned
using Siemens SOMATOM Plus 4 Power system in Medical Imaging Laboratory of
CMPICSU Research Centre in Politehnica University of Timisoara.

The complete scanning of the vertebra includes a number of 67 images provided
in DICOM format and 1 tomogram. DICOM is a standard format developed for
storing, handling, printing, and transmitting the medical imaging data [17, 42]. The
sequential scanning method for image acquisition on dorsal–ventral scanning direction with a resolution of 1 mm was used.

Some representative grayscale images obtained by scanning are presented in
Fig. 7.1, accordingly to three view directions: dorsal–ventral, left–right, and cranio–
caudal. In each image, the light areas (white) represent high-density bone, while
dark areas represent low-density tissues or free space (black).

The quality of the 2D images can be evaluated by the amount of noise and artifacts in the images. In our case the unwanted elements in the images are few, so the
quality of image acquisition was high. When noise and artifacts occur, the appearance of the surroundings contains small white pixels. The isolated gray pixels inside
the vertebral contour must not be confused with noise; they represent the structure
of the spongy bone.

In raw images, linear and angular measurements can be performed in order to
establish the geometrical characteristics of the facet joints. This anthropometrical
data are used in the equilibrium study to establish the load directions.

In Fig. 7.2 are presented two examples of slope measurements on the transversal
facet joint, in two planes: frontal and lateral. The angles of the facet joint measured
in lateral plane respecting the horizontal direction differ from top to bottom site
with about $3.6°$. This difference is an anatomical characteristic which allows angular positioning of two adjacent vertebrae and obtaining in this way the physiological
cervical curvature.

7.2.2 Image Processing

In image computing, a grayscale digital image is an image where the value of each
pixel is a single sample that carries out all the information about the intensity.

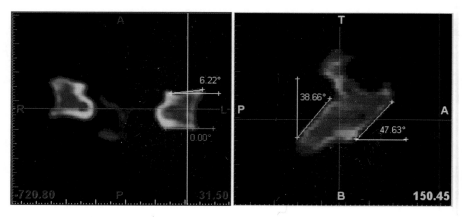

Fig. 7.2 Angles of the facet joints in frontal and lateral planes

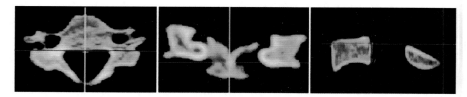

Fig. 7.3 Overall color mask applied on slices

These kinds of images are composed exclusively of shades of neutral gray, varying from black at the weakest intensity to white at the strongest one [2, 24]. In bone tissue imaging, the dark gray represents the low-density bone tissue and the range from light gray to white corresponds to the cortical bone tissue.

The image quality was evaluated from both background and areas of interest point of views. A desirable background is in dark gray tone or black without intensity variations, noise, or artifacts. In this case, the lack of disturbing elements leads to a high-quality background. The areas of interest have good defined contours toward the background.

The main purpose of the image processing was to convert the discrete 2D images into a 3D volume. In order to make this possible all images were imported into *Mimics* image-processing software. The volume conversion protocol uses several steps of masking the areas of interest. The main characteristic of one mask is the thresholding interval, which is the simplest method of image segmentation [24, 38]. Using this function, individual pixels in a grayscale image are marked as "object" pixels if their value is greater than the set threshold value. Typically, an object pixel has a value of "1" while a background pixel has a value of "0."

Taking into account that the purpose was to create a multi-solid 3D model (a model composed of the inner body and outer shell), two different masks were necessary to be applied. The green mask represents the first mask which was applied over the entire area of the scanned vertebra, like in Fig. 7.3.

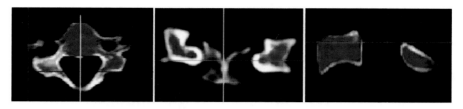

Fig. 7.4 Selective color mask applied on slices

In this case the differences in bone density were not taken into account. The threshold interval was set between −824 and 2,050 Hounsfield units (HU). The negative limit of the threshold corresponds to the dark gray tone, while the positive value corresponds to the light gray.

The threshold limits were set very close to the interval limits (−1,024 to 2,050) due to the large variation of the gray intensity within the image.

The second mask only covers the spongy areas of the bone (see Fig. 7.4). It was applied manually and selectively, only over the dark gray areas, which represent the low-density bone. The automatic threshold interval was set between −824 and 125. The positive limit is so low because it corresponds to the cortical bone which must not be part of the reconstruction. Using only the automatic mask generation was not satisfactory in this case because the threshold interval includes also the values of transition tones between vertebra's contours and background. The quality of the resulted model is directly influenced by the mask application together with the threshold interval.

Having the two color masks, one for the inner area of the bone and a global one, two reconstructions were performed. The transformation of a 2D set of masks into a single volume was made by contour interpolation. The contour interpolation is a 2D interpolation in the plane of the images that are smoothly expanded in the third dimension. This interpolation algorithm uses the gray value interpolation within the slices, and a linear interpolation between the contours in Z direction. The interpolation methods give enough accurate results for the modeling purpose.

The volumes were computed and exported as a point cloud format. This format contains the x, y, and z coordinates of the points that describe the two volumes in the Cartesian space (see Fig. 7.5).

For many medical applications, an exact segmentation of the vertebral column including an identification of each vertebra is essential. However, although bone structures show high contrast in CT images, the segmentation of individual vertebrae is challenging [32].

7.2.3   Surface Generation

The beginning of *surface phase* deals with two point cloud files of both inner and outer vertebral volumes. The meaning of this phase is to create a certain number of

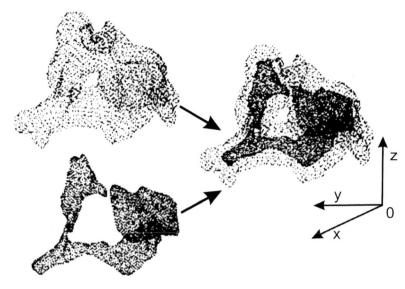

Fig. 7.5 Inner, outer, and assembly parts of the vertebral volume, in point cloud representation

complex shape patches, from the given point clouds. The number of patches represents a compromise between the fidelity of the original model and the maneuverability in simulations of the resulted model. By enclosing the multitude of patches, a shell structure was built. The tight closed volume can be saved as a solid model.

The conversions of the point cloud formats to surfaces were made into the Geomagic Studio environment [12, 21].

The outer point cloud contains an amount of 6,694 points, while the inner point cloud contains a smaller number of 4,642 points. At this stage, some isolated points were selectively eliminated from the cloud in order to prevent a surface generation failure. The selection of the erasable points was made taking into account the native geometrical imperfections of the vertebra.

During the wrapping process, the two point clouds were transformed apart in two polygonal shells. The outer volume consists of 13,374 triangular surfaces while the inner volume consists of 9,284 triangles.

The mesh structures of both objects prove to be very rough after initial wrapping (see Fig. 7.6), a refinement of the element size being demanded. To achieve smooth meshes, the bed triangular elements were first repaired and reduced eight times in size after. The smooth structures (see Fig. 7.6) are now ready to be transformed into *shape phase*.

The next stage of surface generation was to replace the fine mesh with larger patches (NURBS surfaces) in order to generate a more handling object with the same geometrical characteristics (see Fig. 7.7).

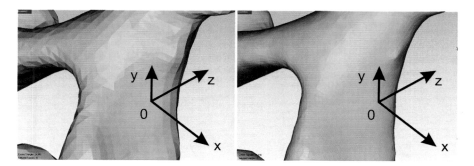

Fig. 7.6 Rough and smooth polygonal meshes of the cervical vertebra

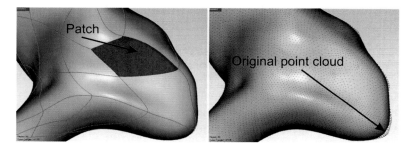

Fig. 7.7 Patched surfaces and loss of geometrical fidelity while constructing the object

In order to perform this transformation in Geomagic Studio environment, the following steps have to be accomplished:

- Computing the contour lines
- Extending the tangent lines
- Filling all the existent gaps in the mesh
- Computing the grid
- Repairing the grid intersections if needed
- Constructing the NURBS surfaces [21, 45]

The surface generation phase ends by exporting the models as IGES format files.

7.2.4 Solid Generation

The IGES format files were imported into SolidWorks environment for further processing. IGES is a standard CAD format for solid objects, designed to work with the majority of CAD environments. This was the reason why this format type was chosen.

When diagnostic tool runs in SolidWorks, some of the surfaces are detected as faulty or gaps. A faulty surface is sometimes generated by self-intersecting patches.

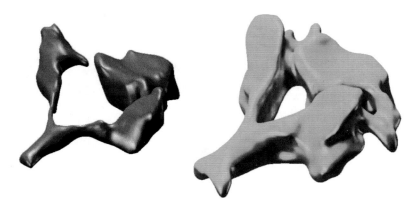

Fig. 7.8 3D views of the core and shell bodies

The error number is usually proportional with several factors: the native number of points, smoothness of the mesh, geometrical dimensions, and complexity of the geometry. Due to these factors, the inner solid experiences more errors than the outer solid.

Using the surface module of the software, the inadequate surfaces were erased or modified. The manual procedure consists of the following steps:

- Erasing the inadequate surfaces
- Drawing a 3D spline to divide a large inadequate surface into smaller good ones
- Replenishing the empty surfaces

This task can also be automatically done but with less chances of success, the simple vertex repositioning being not enough. When all the surfaces from both volumes were enclosed, the conversions into solid bodies were automatically made.

In the next stage the two solids were combined in one part by matting their origins as coincident. Using Boolean subtraction method, the inner body (core) was eliminated from the outer solid, having as result a new shell body. In this way, the core and the shell bodies were created (see Fig. 7.8).

The reason of having two solids rather than one is obvious: the core corresponds to the spongy vertebral bone, and will get the spongy bone properties, whereas the shell corresponds to the cortical bone, and will get suitable properties.

The desired multi-solid body represents the assembly of the core and shell solids. For relative positioning of the bodies the origins were set as coincident. This operation was possible because the origin of the parts is identical. The coincidence of the origins is derived from the reconstruction stage, where the point clouds were exported from the same processed images. The multi-solid body of the cervical vertebra is presented in Fig. 7.9.

The reconstruction stages of the cervical vertebra can be extended to any elements of the human body. The required conversion stages from CT images to solid objects can be synthetically presented as a transformation protocol (see Fig. 7.10).

Fig. 7.9 3D views of the multi-solid vertebral body in transversal section, frontal section, sagital section, and isometric view

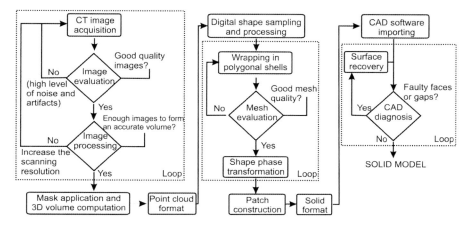

Fig. 7.10 Image to solid transformation protocol

The protocol indicates a successful way to generate a complex shape solid body. The protocol can be applied to any scanned images in order to obtain solids.

The cervical vertebra presented in Fig. 7.9 can be used to model the functional unit of the cervical spine. Additionally, soft tissues like facet joints, ligaments, muscles, spinal cord, and nerves can be integrated into the structure in order to achieve a more realistic model.

The soft tissues of the cervical unit have not been part of the reconstruction process. Elements like the intervertebral disc, facet joints, interspinous ligament, spinal cord, and nerves were modeled using classical CAD tools. The reconstruction of these tissues can be possible by MRI scanning of a living cervical spine, the CT scanning being not adequate for soft tissue reconstruction.

The reconstructed elements together with the modeled ones were assembled in one compact structure which is presented in Fig. 7.11, in posterior and anterior views. The muscle development was not yet modeled, representing a further work.

The relations between elements are coincidence, tangency, concentricity, parallelism, and distances.

CAD environment offers data about the properties of the multi-solid bodies such as mass, volume, moments of inertia, and mass center. The existence of these properties proves the availability of the model for mechanical simulation tests.

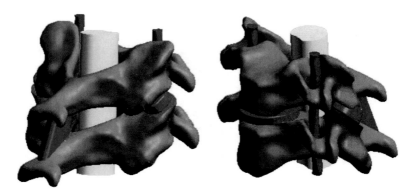

Fig. 7.11 Isometric-posterior and -anterior views of the reconstructed cervical functional unit

7.3 Classical Geometric Modeling of a Functional Implanted Unit C2–C3

Due to the complex structure and biomechanical behavior of the human spine there are many factors determining the spine injuries. Benzel [5] describes the injury types based on the mechanism of injury. For example, pressure on the spinal cord in the cervical region can cause serious problems because virtually all of the nerves to the rest of the body have to pass through the neck to reach their final destination (arms, chest, abdomen, legs). Thus, the goal of cervical spine surgery is to relieve pain, numbness, tingling, and weakness, restore nerve function, and stop or prevent abnormal motion of the spine [5].

Metallic plates, screws, or wires are often used in order to stabilize the cervical spine. The implants are used to facilitate fusion, correct deformities, and stabilize and strengthen the spine. Most spinal implants are made of biocompatible materials: metals such as titanium, titanium alloy, or stainless steel; some are made of nonmetallic compounds such as polymers and biodegradable polymers. The implants have many different shapes and sizes in order to accommodate patients of all ages. These are manufactured to conform to the topology of the spine and are held in place by screws set into adjacent vertebrae. When the plate requires adjustment, a contouring tool is used to customize the fit to the patient's anatomy.

There are many factors to be considered in the design of spinal implants. Each of these factors must be adequately treated in order to obtain an optimal result [5]. Thus, the model of the functional implanted unit has to be developed and analyzed.

Because of the high computing requirements needed when simulating the biomechanical behavior of the reconstructed functional unit of the cervical spine, usually a functional model can be achieved using classical modeling technique. This CAD modeling has several advantages than the reconstructed model:

- Easy to constrain the elements in the assembly
- Good anatomical functionality

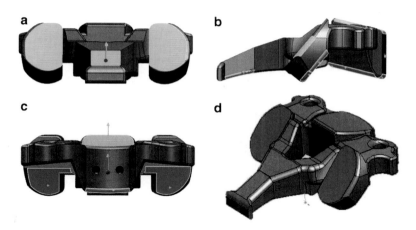

Fig. 7.12 3D views of the constructed C2 vertebra: (**a**) frontal view, (**b**) lateral view, (**c**) posterior view, (**d**) isometric view

- Highly maneuverable
- Small file size
- Can be adapted and modified at any time

Due to its advantages, the model can be further more developed by attaching an anterior stabilization plate on it.

The geometrical modeling of the functional implanted unit involves two different aspects of modeling: the design of the anatomical elements and design of the mechanical elements. Using *Feature Based Design* module of the SolidWorks environment, the anatomical and mechanical elements were modeled [3, 10, 35, 50].

The model of the functional unit of the cervical spine is composed of the following basic elements: two vertebrae (C2 and C3 in our case), intervertebral disc, interspinous ligament, two facet joints, spinal cord, and vertebral arteries. Certainly, the natural functional unit is far more complex than that, but the proposed functional unit successfully fulfills the shape and functioning requirements with the mentioned elements.

To accomplish the functional requirements of the cervical unit, a great attention was paid in building the shape details of each element, and the contact surfaces between the various elements.

The modeling complexity is proved by the number of operations on the vertebra, as many as 70 for each vertebra. In design process, operations like extruded boss, extruded cut, revolved cut, swept boss, fillet, chamfer, simple hole, and loft were used, starting from various primitive sketches. The modeling concept was based on *extrude cutting* of different contours, from a 3D block.

The result of the primitive sketch subtractions, together with many *feature* operations, leads to a C2 vertebral body having the shape and functional surfaces similarly to the natural one. In Fig. 7.12, different views of vertebral body are presented.

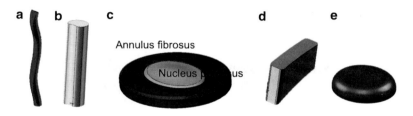

Fig. 7.13 Constructed soft elements of the cervical functional unit: (**a**) vertebral artery, (**b**) spinal cord, (**c**) intervertebral disc, (**d**) interspinous ligament, (**e**) transversal facet joint

The highlighted areas represent the functional surfaces which are designed to join with the anatomical soft elements.

In addition to the mentioned features, each vertebral body received two threaded holes designed to join with the bone screws while assembling the functional unit. Threaded holes design is considered as secondary preparatory operation.

The soft tissues which connect the vertebral bodies were modeled by positive extrusion method (*boss extrusion*) of various primitive sketches. Excepting the intervertebral disc, each soft element (vertebral artery, spinal cord, interspinous ligament, and transversal facet joint) is represented by single one component having its proper shape (see Fig. 7.13a, b, d, e) and features.

The intervertebral disc is an exception due to its different properties in external and internal regions. The *annulus fibrosus* structure was modeled as an ellipsoidal ring grew linear in the Z direction, while the *nucleus pulposus* structure as an ellipsoid body. The two disc's elements were assembled using both coincidence and concentricity mates (see Fig. 7.13c).

Traumatic cervical spine injuries are successfully stabilized with bone plates applied to the anterior vertebral bodies. Various anterior cervical plate systems are currently available on the market, such as Caspar Plate and ABC (Aesculap), Synthes Plate, Codman Plate (Slim Lock) (Johnson and Johnson), Atlantis plate (Sofamor Danek) [23, 62], SC-AcuFix® Anterior Cervical Plates [69], MaxAn® Anterior Cervical Plate System, C-Tek® Anterior Cervical Plate System [6], Reflex Hybrid ACP Implant [52], Cervical Spine Locking Plate (CSLP), and Vectra-Vectra T [57].

The implant was designed to offer anterior fixation possibilities for varying anatomies and pathologies. The modeled bone plate is a small implant designed to join two consecutive vertebrae. There are also many other types of bone plates designed for three or more vertebrae stabilization.

The mechanical purpose of the stabilization implant is to create a rigid fixation between the C2 and C3 vertebrae. The stabilization system integrates the following elements: one bone plate coded CP1, four bone screws of three types coded *Sr1*, *Sr2*, and *Sr3*, and four identical locking screws.

The cervical bone plate was designed using *feature design* module. The primitive sketch was constructed in the top plane of the working environment. The basic *bone shape* of the plate was achieved by filleting all corners and using circle arches.

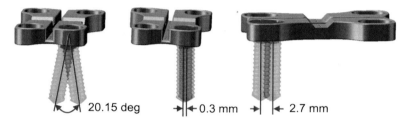

Fig. 7.14   Bone screws positioning ranges with respect to the bone plate

Since the development trends in cervical bone plates are increasingly turning to more dynamic assembly type, the modeled system presents two major features:

• Capability of angular positioning and mobility within an important interval of the bone screws
• Possibility of longitudinal and transverse migration of bone screws along the bone plate axes

These two functional characteristics are coming to eliminate the stress shielding effect produced by the rigid fixation to the bone graft. The large angular positioning capability provides an appropriate flexibility for intra-operatory fixation of the implant. It also allows an appropriate placing of the plate during the osteointegration. The longitudinal and transverse migration capacity assures adapting of the relative position of the plate to the bone structure. The allowed migrations must be within certain limits and developed under the action of loads greater than gravitational force generated by own masses.

The intra-operative bending capacity of the plate is sustained by its geometrical configuration together with the longitudinal and transversal grooves. This design facilitates bending around longitudinal and transversal axes, and suites with anatomical shape of the vertebral body. The four mounting holes have large dimensions and elliptical shape to allow appropriate angular, longitudinal, and transverse positioning of the bone screws. The nine fine holes having the diameter of 1.5 mm are necessary to temporarily position the plate in implantation site, before the final fixing. The connection radius is 3.5 mm facilitating a possible longitudinal bending of the plate.

The relative positioning between the bone plate and bone screws can be seen in Fig. 7.14, where for a better visualization the screws were declared transparent. The angular positioning of the bone screw in front plane ranges from 0 to a maximum of ±10°. This means that the angle of convergence of two bone screws on the same line is 20°, which defines an appropriate triangulation effect. A wider angular range could be obtained by reducing the screw head or plate's height. Figure 7.14 shows also the allowed relative translation of the screws along the transversal axis of the plate. This value although very small (0.3 mm) produces a significant mobility of the bone plate with respect to the vertebrae. A third possibility is the large longitudinal movement (see Fig. 7.14). The value of this displacement is of 2.7 mm for a screw.

The bone screws have three different designs. Having the same diameter and length, the differences between the bone screw types consist in the thread's characteristics and screw heads. The threads of the bone screws differ in pitch, helix's angle, and profile's shape. The three types of developed profiles were coded *Pr1*, *Pr2*, and *Pr3* and materialize threads designed to assemble with cortical bone (*Pr1*), spongy bone (*Pr2*), and a combination of cortical-spongy (*Pr3*) called hybrid. None of these profiles is standardized, they being designed to perform numerical and practical simulations to analyze the functionality of the backbone. The threads of the bone screws were constructed by sweeping the adequate profiles along a helix curve. The mechanical characteristics of the bone screws are presented in Table 7.1.

The correspondences between bone screw types and the type of the bone tissues are the following: the *Sr1* screw fits the requirements of self-tapping and assembling with the cortical bone tissue; the second type *Sr2* has design adapted to interfacing the spongy bone; and *Sr3* is a hybrid solution developed as a combination of the first two.

Since the vertebral body is a spongy core within a cortical shell, the hybrid screw solves the deficiencies of the very specific first two screws. These deficiencies exhibit when a spongy screw for example is penetrating the thin cortical layer, producing micro-cracks due to the large depth of the thread. On the other hand, when a cortical screw progresses into the spongy bone, the fine thread of the screw generates a very poor assembling.

Sr1 bone screw (see Fig. 7.15a) has a cylindrical head with a rounded bottom edge to materialize a spherical cap. This spherical cap is an essential element in the screw ability to position and tighten at various angular values. The screw tapping is performed with a screwdriver, introduced into the hexagonal socket with an aperture of 1.5 mm.

Sr2 bone screw (see Fig. 7.15b) has a "ball" head made by combining two different spherical caps differentiated by different radii. The lower spherical cap, with a radius greater than *Sr1*, allows screw placement within a larger angle range, making the screw more flexible in terms of implantation possibilities.

Sr3 bone screw (see Fig. 7.15c) is completely different from constructive point of view related to the others. In addition to hybrid type of adopted profile, it was provided with a particular head. It consists of a *flower* type element which is deformed by a second screw when these two are assembling. The expanding of the *flower* has the purpose to achieve a contact between the screw head and the hole of the plate. The bone screw *Sr3* was developed to achieve fixation by one cortical breakthrough, so it is enrolling in those of unicortical type.

In Fig. 7.15 the models of the bone screws are presented together with the bone volumes. The bone volumes represent the tissue volumes engaged between threads, along the entire length of the screw. These are active volumes which hold the screws in place when a pullout force is applied on the screw. So, higher volumes mean higher pullout resistance for the screw.

However, in our case, the bone volume corresponding to the hybrid bone screw will prove to be high enough for a good mechanical strength. This bone screw

Table 7.1 Screw characteristics

Screw type	Optimal implantation site	Thread pitch (mm)	Thread height (mm)	Helix angle β (°)	Helix diameter (mm)	Head type	Bone volume (mm³)
$Sr1$	Cortical	0.68	9.06	13.53	2.9	Cylindrical	9.3
$Sr2$	Spongy	0.70	9.06	14.53	2.9	Spherical	20.7
$Sr3$	Hybrid	0.60	9.06	12.17	2.9	Expandable	12.7

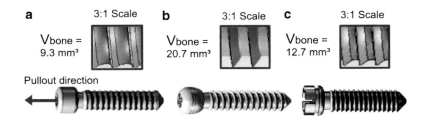

Fig. 7.15 Screw models and the bone volumes engaged between the threads: (**a**) Sr1 bone screw, (**b**) Sr2 bone screw, (**c**) Sr3 bone screw

Fig. 7.16 Exploded view of the implanted cervical unit

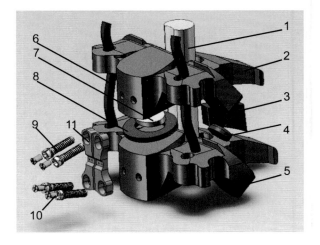

represents a compromise between the pullout resistance and the micro-crack prevention. The design concept of *Sr3* bone screw requires an additional locking screw. This has the purpose of locking the fixation screw into the elliptical hole of the cervical plate, by expanding the *flower* head of the fixation screw. The locking screws were designed to be used only with the bone screw *Sr3*, and their purpose is to hold in place the bone screws, and prevent unscrewing, by expanding their heads against the cervical bone plate. The locking screw is screwing into the bone screw and, due to its tapered head, it creates a wedge effect. The *Sr1* and *Sr2* screws have no locking system into the bone plate and therefore can experience unscrewing during the movement of the spine.

Having the models of all anatomical and mechanical parts of the cervical implanted unit, these were assembled into a SolidWorks assembly file. In order to create a valid assembly, constraints as coincidence, concentricity, parallelism, distance, and so on were used. These constraints allow to manipulate the model as it behaves in reality, excepting the elastic properties of the soft tissue (all the parts are treated as rigid bodies in the assembly environment). The exploded view of the assembly is presented in Fig. 7.16, having all the components visible: 1—spinal cord, 2—C2 vertebra, 3—interspinous ligament, 4—transversal facet joint, 5—C3 vertebra, 6—vertebral artery, 7—nucleus pulposus, 8—annulus fibrosus, 9—bone screw, 10—locking screw, and 11—bone plate.

At this stage we can say that the assembly fully fulfills the requirements imposed in the beginning, those of functionality and geometrical similarity with the physiological structure.

The modeling and assembling techniques could be successfully used for the development of a functional implanted unit at any spinal level.

7.4 Simulation of Mechanical Behavior for ICU

The multitude of the complex shape elements involved in cervical functional unit, implanted or not, makes the analytical analysis of the structure's biomechanical behavior difficult. By applying such levels of simplifications in order to make possible a classical approach, the behavior of the structure can dramatically differ. So, the FEM is more suitable for our simulation purpose.

FEA is a method developed to compute a real model in a mathematical form for a better understanding of a highly complex problem [1, 19, 33]. The geometric structure is defined as a finite number of simple elements composing a mesh. More finite elements in a structure mean higher accuracy of the analysis.

Having the model of the ICU developed in Chap. 3, this chapter proposes a mechanical simulation of this functional unit, using *static structural* module of the ANSYS Workbench environment.

The study's main objective is to identify the stress–strain conditions at the bone–metal interface for certain loading and restricting conditions. The bone–metal interface is defined by the contact surface between the screw threads and the bone engaged between the threads. There is also a metal-to-metal interface materialized by the contact surfaces between the screw head and the implant plate, which is not part of the study.

We simulated three cases, respecting the three different geometries of the bone screws *Sr1, Sr2,* and *Sr3.* The studies were individually performed for each implanted unit, by following the same steps and under the same loading and fixation conditions.

For valid simulation results, parameters like precise contacts between parts, material properties, mesh size, and loading and fixing conditions were set up for ICU and presented in the following.

7.4.1 FEA Main Stages

In order to perform the numerical analysis of the ICU, the ICU assembly was exported from CAD environment in *.sat* file format. The individual connections between the parts represent a feature of the assembly, which is transferred to the

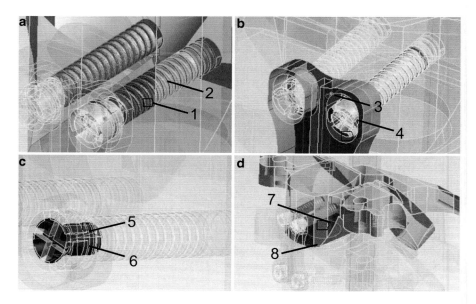

Fig. 7.17 Defining of contacts between ICU elements: (**a**) contact between bone screw 1 and bone 2, (**b**) contact of bone screw head 3 and bone plate 4, (**c**) contact between bone screw 5 and locking screw 6, (**d**) contact between vertebra 7 and intervertebral disc 8

.sat file. But in some cases, when the CAD file is imported into the simulation environment, some valid contacts may miss, while unwanted others can be present. So, for simulation rigor, the mechanical connections between the assembly's elements were manually declared. Overall, the connections between the intersecting elements were materialized into 24 bonded type connections.

Defining of accurate contacts represents a very important factor in stress–strain evaluation [50]. The contacts between the bone screw threads and the conjugated bone interface (see Fig. 7.17a) were set as bounded in order to simulate a physical contact behavior. In the same way, the contacts between the bone screw head and the bone plate (see Fig. 7.17b), and between the bone screw and locking screw (see Fig. 7.17c), were established. The contacts between vertebrae were declared through the elastic elements: intervertebral disc, facet joints, and interspinous ligament (see Fig. 7.17d).

The material assignation represents a mandatory condition for static structural analysis. In order to achieve valid results, the materials declared in the simulation have to assure the same mechanical properties as the natural ones. In the simulation environment some material types were presented in the database, while others (bone, disc, and ligament) were declared accordingly to the literature [13, 20, 68].

The materials selected for the structure's components are:

- Spongy bone for the internal vertebral bodies
- Titanium alloy Ti-6Al-4V for bone screws and cervical bone plate
- Stainless steel for locking screws

Table 7.2 Properties of the materials of ICU components

Material	Young's modulus (MPa)	Poisson ratio (–)
Titanium alloy	110,000	0.280
Stainless steel	193,000	0.300
Cortical bone	12,000	0.300
Spongy bone	100	0.200
Ligament	18	0.350
Annulus fibrosus	450	0.300
Nucleus pulposus	0.13	0.499

- Ligament properties for the facet joints and interspinous ligament
- Their own characteristics for the nucleus pulposus and annulus fibrosus

In static structural analysis, material properties like Young's modulus and Poisson ratio are essentials. Their values are presented in Table 7.2 for each material involved in the study, as a result of a bibliographic selection [13, 20, 68]. The properties presented in Table 7.2 are essentials and sufficient for this type of evaluation [34].

The mesh structure generation using tetrahedron elements can be created in two ways: in automatic mode and by setting up the element size separately for each component [39]. In the automatic mode, the element size is modified in order to adjust the geometrical requirements of the components. In order to have a constant fine mesh accordingly to the part's sizes, a manual mesh refinement was used.

In our study, the mesh characteristics are a constant element size of 1 mm for the mechanical parts and another mesh structure of 2 mm element size for the anatomical parts (see Fig. 7.18). The reason of choosing a smaller element size for the mechanical elements results from their reduced dimensions, comparing with the anatomical structure.

Defining the same small size elements for the whole assembly will dramatically increase the computing requirements due to the complex geometry of the vertebrae.

The loading conditions applied to the structure were assigned taking into account the gravity force generated by the head and neck's masses, and the bending forces and moment which act in a normal flexion movement.

The gravity effect acts as distributed forces on three surfaces: the vertebral body plateau and symmetrical facet joints. The quantum of these forces usually represents 5–7 % of the whole body weight [30], so for a subject having 70 kg weight, the gravitational force at C2 vertebral level is around 50 N. This value is distributed (F_1, F_4, and F_5) on the three surfaces: vertebral body (vertical force $F_1 = 30$ N) and the symmetrical facet joints. The forces on the facet joints have values of 10 N each, resulting in a value of perpendicular component force of 6.8 N. In simulation, only the perpendicular component on the facet joint was considered. This has a value diminished by the facet's angle (see Table 7.3).

The bending movement of the head was simulated by a bending moment M_f applied perpendicular to the spinal axis and two additional shear forces F_2 and F_3 simulating the pullout effect on the bone screws. The bending moment M_f of 2.5 Nm

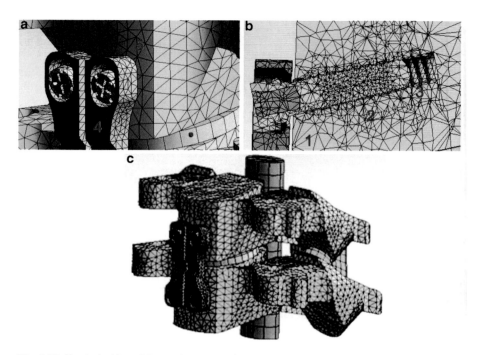

Fig. 7.18 Particularities of the mesh structure for ICU: (**a**) mesh size 1 mm for the bone plate 4 and 2 mm for the vertebra 5; (**b**) mesh size transition in vertebral bone: large mesh 1, transition elements 2, and fine mesh 3; (**c**) overall mesh structure of the ICU

Table 7.3 Loading conditions for ICU, in flexion movement

Loading type	F_1—A spot (N)	F_2—B spot (N)	F_3—C spot (N)	M_f—D spot (Nm)	F_4—E spot (N)	F_5—F spot (N)
Value	30	40	40	2.5	6.8	6.8

was applied on the C2 vertebral body having the direction of forward rotation, and representing the average action of the deep cervical flexors [43]. Kurtz [34] estimated the individual values of these shear forces F_2 and F_3 to 50 N. But, taking into account the bending moment applied on the C2 vertebra, the F_2 and F_3 forces were reduced to 40 N each.

The distribution of the forces and moment on the ICU structure can be observed in Fig. 7.19, together with the fixing conditions. The fixed surfaces are considered the bottom joint faces of the C3 vertebra. Without minimum one fixed support, the analysis cannot be completed.

In addition to the parameters mentioned before, the simulation environment requires some evaluation parameters like stress, strain, or deformation.

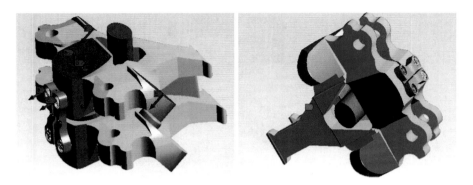

Fig. 7.19 Loading and fixing conditions applied on the ICU structure

7.4.2 Mechanical Effects at the Bone–Screw Interfaces

The bone–screw interface is the boundary of two materials: the vertebral bone and titanium alloy. The crucial differences between these materials are the Young's modulus, the anisotropy of the bone comparing with the titanium screws, and the bone's inhomogeneity.

In the performed simulation, simplifications like isotropy and homogeneity of the bone' material have been assumed. Since the titanium has a higher Young's modulus than the bone (see Table 7.2), from the theoretical point of view, higher stress concentration in the bone screws and higher strains in the bone were expected. This theoretical principle was sustained by the results of simulation.

The stress and strain records that are presented were sampled from screw and bone sections and they refer to equivalent elastic stress and strain according to the *von Misses* yield criterion. These were computed by the simulation software using (7.1) and (7.2) formulas [19, 33], where v' represents the effective Poisson's ratio:

$$\sigma_{\text{Von}-\text{Mises}} = \left(\frac{1}{2}[(\sigma_1 - \sigma_2)^2 + (\sigma_2 - \sigma_3)^2 + (\sigma_3 - \sigma_1)^2] \right)^{1/2} \qquad (7.1)$$

$$\sigma_{\text{Von}-\text{Mises}} = \frac{1}{1+v'} \left(\frac{1}{2}[(\varepsilon_1 - \varepsilon_2)^2 + (\varepsilon_2 - \varepsilon_3)^2 + (\varepsilon_3 - \varepsilon_1)^2] \right)^{1/2} \qquad (7.2)$$

The first aspect of this simulation is focused on the strains acting at the bone–screw interface. In bone–strain evaluation the challenge was to establish whether the bone can support or not the strain action, and which is the strain distribution along the bone threads. Due to the symmetry in lateral plane of both screws' positioning and loading conditions, the results were evaluated and presented for one side only. The behavior of the superior screw Ss (tapped into the C2 vertebra) and the inferior screw Si (tapped into the C3 vertebra) is highlighted.

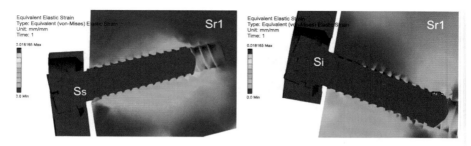

Fig. 7.20 Bone strains acting on Sr1 superior and inferior mountings

The first simulation was performed for *Sr1* bone screw mounting. In Fig. 7.20, the results are presented as color maps, where the color spectrum ranges from dark blue to bright red, having the higher strain values in red color.

By comparing the strains acting on the bones at the Ss and Si interfaces, different effects are identified. The differences in strains show us how the loads acting on the C2 vertebra are transmitted to the bones, bone screws, and bone plate. Since the superior bone screws (axially loaded) manifest tension in the bone interface, the inferior bone screws manifest an indirect tension in the lower longitudinal section of the tapped hole. The tension borne in the lower screw is caused by the bending effect of the plate.

In a pure tension test, the strains should be uniformly distributed on the threads. But, due to the flexion moment applied to the structure, and also due to the gravitational forces, the strain distributions on holes are anything but uniform. In the superior hole, the strains are higher next to the lower section of the screw tip, while in the inferior hole, the strains are higher next to the upper section of the screw tip.

This coupling effect is manifested in both bone screws and it generates a compression effect on the bones. This strain variation is graphically presented in Fig. 7.21, as function of threads number. From the biomechanical point of view, this phenomenon leads to local bone micro-cracks, and sometimes can cause failure. The bone failure represents the first step in screw pullout effect, and the beginning of the implant system failure.

The subject of the second simulation was the bone screw *Sr2*, and its effect on the bone. In order to acquire comparable results between screws effects on the bone structure, the loading and fixing conditions of the simulation were maintained. Also, the position order of the screw and the contact arrangements are identical. The results in Fig. 7.22 reveal an identical strain distribution as in the first case, which proves the repeatability of the simulation. The differences appear in the strain values. The maximum value of strains in the superior section is around 0.004 mm/mm meaning only a half of the value recorded in the first case.

As the variation graphics indicate (see Fig. 7.23), the lower strain induced by the bone screw *Sr2* into the bone interface can be directly influenced by the thread's profile. This proves that a deeper thread, which engages more bone between the

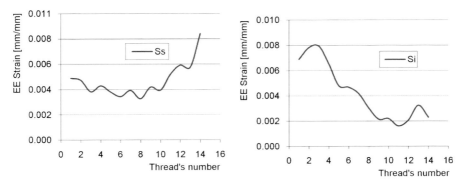

Fig. 7.21 Strain variation along the bone threads for Sr1 superior and inferior mountings

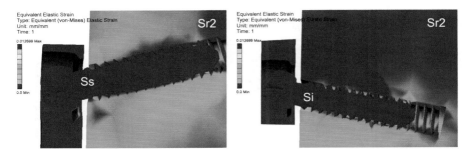

Fig. 7.22 Bone strains on Sr2 superior and inferior mountings

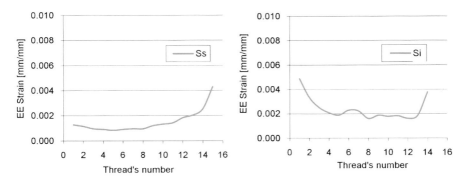

Fig. 7.23 Strain variation along the bone threads for Sr2 superior and inferior mountings

threads, will distribute the strains to a larger bone volume. The biomechanical effect has a very good score here because of the pullout strength enhancement. The disadvantages in this case are caused by the poor tapping characteristics of this screw. It behaves well in cancellous bone, but generates cracks when penetrating the cortical layers.

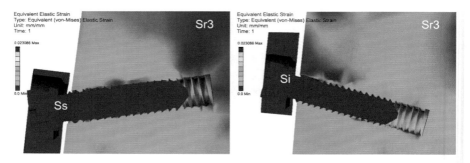

Fig. 7.24 Strain variation along the bone threads for Sr3 superior and inferior mountings

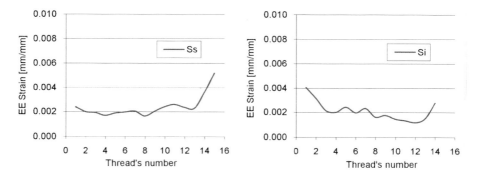

Fig. 7.25 Strain variation along the bone threads for Sr3 superior and inferior mountings

The third set of simulation results put on evidence the *Sr3* hybrid screw behavior. The values of the strains distributed along the superior thread hole (see Fig. 7.24) are sensibly higher than in the *Sr2* case, but smaller than in the *Sr1* case. In the inferior section of the hole, the strains almost overlap with the second case results.

As Fig. 7.25 presents, the strain variations along the threads are less present in this case. This indicates us a better strain distribution on the bone interface. This screw type represents a geometrical compromise of the thread's depth, pitch, and profile, and proves to represent a good compromise from the strain distribution point of view.

The ability of strain distribution in bones is better highlighted by representing the performances of the three screws on the same graph (see Fig. 7.26). Here are presented the behaviors of the superior bone interfaces for each screw mounting, and inferior bone interfaces.

The second aspect of this simulation is focused on the stresses acting at the bone–screw interface. Because of the high mechanical strength of the bone screws, they exhibit larger stresses than the bone structure, making no difference of screw

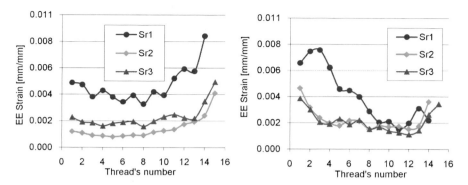

Fig. 7.26 Strain effects on bone, induced by Sr1, Sr2, and Sr3 superior and inferior screws

type. The stresses were sampled along the screw sections, in the thread's adjacency. This area usually represents a stress concentration due to geometrical leaps.

In the case of *Sr1* bone screw, the stress values decrease from a maximum of 22 MPa. The nonsymmetrical distribution toward the geometrical axis of symmetry can be observed in Fig. 7.27a, b. This behavior is generated by the complex loading of the structure and is leading to bone strain distribution.

The *Sr2* bone screw stress conditions are similar to the *Sr1* bone screw. The maximum stress recorded was 30 MPa, with a descendent trend from head to tip. The high stresses recorded (see Fig. 7.27c, d) are caused by the threading depth of the screw. In the stress distribution, an important role is played by the contact surface between the screw head and the plate, this surface being responsible to load sharing. So, the uniform distribution aspect of the stresses for the superior and inferior *Sr2* bone screws can be caused by the spherical screw head.

The maximum stress value for *Sr3* bone screw was 35 MPa, recorded in the proximity of the screw head (see Fig. 7.27e, f). The stress decreases from the head to the tip of the screw. For the inferior screw, a uniform stress distribution can be observed along the screw and also toward the symmetry axis. This effect can be assigned to the thread profile and the rounded contact between the screw head and bone plate.

Better stress distribution in bone screws leads to good strain distribution in bone structure.

The stress tendencies for all the bone screws in superior and inferior vertebral assembling are presented in Fig. 7.28. For each screw in superior assembling, a descendent tendency from the head to the tip can be observed. This stress is generated by the pullout loadings, applied directly on the screws. For the inferior screw, the effect is attenuated due to the complex load sharing. The load sharing effect in inferior screws is related to the contact and relative positioning between the bone screw heads and bone plate.

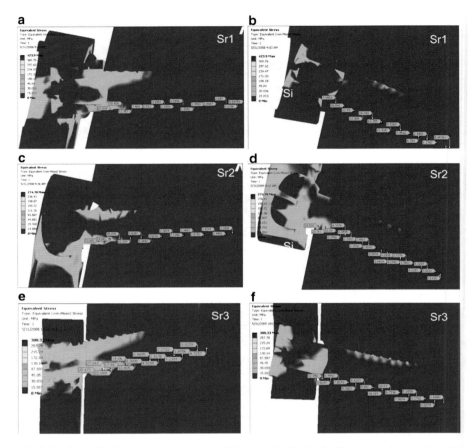

Fig. 7.27 Stress distribution in bone screws: (**a**) stress distribution for Sr1—superior position, (**b**) stress distribution for Sr1—inferior position, (**c**) stress distribution for Sr2—superior position, (**d**) stress distribution for Sr2—inferior position, (**e**) stress distribution for Sr3—superior position, (**f**) stress distribution for Sr3—inferior position

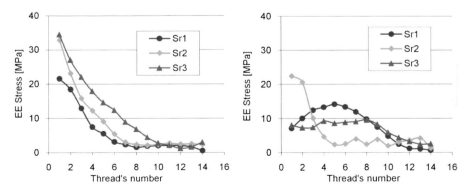

Fig. 7.28 Equivalent elastic stress along the superior and inferior screw threads

7.5   Conclusions

Classical design and solid reconstruction are two modeling techniques aiming essentially the same objective: developing of a realistic 3D virtual model which respects the shape and characteristics of the real object.

The vertebra models obtained by classical design are focused on functional models, sometimes to the detriment of appearance. On the other hand, 3D reconstruction models developed for the same structures are focused on virtual multi-solids, functionally and aesthetically identical with the real ones.

Using classical design, solid models like the vertebrae, intervertebral disc, spinal cord, facet joints, and ligaments were built. The models benefit by several advantages: functionality, small volume of data, facile handling of the model, and shape and dimensions that are appropriate with the target model.

The 3D reconstruction of the vertebrae led to a systematic processing protocol of the images scanned by CT into solids, using two environments: one for image processing and a CAD one. An important objective achieved by 3D reconstruction is the developing of multi-solid models of the vertebrae. For inner core and outer shell structure, different physical and mechanical properties can be assigned.

A synthetic analysis of the differences between the two types of geometric modeling is presented in Table 7.4. It can be seen that the benefits of solid reconstruction in terms of accuracy and functionality actually generate two disadvantages: low changeability and handling. The changeability of the model refers to the object's property of being modified or transformed while handling quantifies the computational resources needed for working with the model.

Conclusively, both classical design and solid reconstruction present specific advantages and disadvantages. It is important to note, however, that choosing one over the other should be fully in line with the requirements imposed by the modeling objective.

In order to achieve a more realistic model, both models will be developed taking into account additionally the soft tissues such as facet joints, ligaments, muscles, spinal cord, vertebral artery, and nerves. The 3D reconstruction of these tissues needs MRI scanning of a cervical spine and the same reconstruction protocol. The classical CAD model will assume more fidelity of the anatomical geometry.

Based on the designed stabilization system, new models of bone plates and bone screws will be developed and analyzed.

The results of the simulation highlight two mechanical aspects: the stresses and strains acting at the bone–screw interface. The behavior of the bone screw–plate

Table 7.4 Comparison of the two types of modeling

Modeling type	Shape accuracy	Dimensional accuracy	Functionality	Changeability	Handling
Classical CAD modeling	+	+++	++	+++	+++
Solid reconstruction	+++	+++	+++	+	+

+, Acceptable; ++, good; +++, excellent

interface is very much dependent on the bone screw heads, because their geometries come in direct contact with the plate, while the bone–screw interface is directly dependent on the thread's profile.

Based on the simulation results, we may give the answers to the following questions:

• Are the bone interfaces able to support the applied loads?
• Which screw type provides better pullout strength?

For the first question, the strains in the bone structures are generated by stresses ranging from 1.2 to 3.5 MPa. Comparing the stress values with the ultimate strength of the spongy bone (115 MPa for tension and 165 MPa for compression [29]), the conclusion is that no interface exceeds the strength limit. Therefore, all bone screws provide safety fixations for the loading and restricting conditions of the functional unit.

Regarding the second question, better pullout strength is provided by the screw *Sr2*, because the strains in bone volume are well distributed. But, taking into account that the vertebral body is not entirely a spongy bone but a composite structure of a spongy core inside a cortical shell, and also the fact that the *Sr2* depth of threading generates micro-cracks in the cortical bone interface, we can affirm that the hybrid screw *Sr3* is more suitable for the fixation purposes. It has a medium thread depth and a rounded profile, so it cannot cause cracks when using properly. In the same time, the strains induced in the bone interface have comparable values with the *Sr2* screw.

This simulation approach makes a small image about the influence of the thread type, depth, pitch, and crossing size of the screws, on the bone tissue fixations.

Another important aspect of the simulation must be focused on the stresses and strains acting at the screw–plate interface. This study can be performed using the same model of the ICU under the same assumptions as in case of the study of stresses and strains acting at the bone–screw interface.

The improved models will be subjected to a numerical analysis in order to highlight the same mechanical aspects related to stresses and strains acting at the bone–screw interface and screw–plate interface, in a more realistic case.

References

1. ANSYS Inc. (2005) ANSYS structural analysis guide. Ansys Release 10.0. SAS IP, Inc., U.S.A. pp 26-35, 185–194
2. Bankman IN (2000) Handbook of medical imaging. Processing and analysis. Academic, New York
3. Benhabib B (2003) Manufacturing-design, production, automation and integration. Marcel Dekker Inc., New York
4. Benzel E, Muehlbauer E, Orrico K (2001) New spin on spine: introducing decade of the spine initiative. JD AANS Bull 10(4):43
5. Benzel EC (2001) Biomechanics of spine stabilization. Thieme, New York
6. Biomet Web site. Cervical products. http://www.biomet.com/spine/products.cfm?pdid=3&majcid=13
7. Boisvert J, Cheriet F, Pennec X, Labelle H, Ayache N (2008) Articulated spine models for 3-D reconstruction from partial radiographic data. IEEE Trans Biomed Eng 55(11):2565–2574

8. Bossart PL, Martz HE, Hollerbach K (1996) Finite element analysis of human joints: image processing and meshing issues. Proc Int Conf Image Process ICIP-96(2):285–288
9. Bozdoc M (2003) The history of CAD. http://mbinfo.mbdesign.net/CAD-History.htm. Accessed 20 Aug 2010
10. Bralla JG (1999) Design for manufacturability handbook, 2nd edn. McGraw-Hill Professional, New York
11. CAD – History, present and future. http://en.wikiversity.org/wiki/CAD_-_History,_Present_and_Future
12. Cameron BM, Manduca A, Robb RA (1995) Surface generation for virtual reality displays with a limited polygonal budget. In: Proceedings of the 1995 international conference on image processing. Washington D.C., pp 23–26
13. Chen SH, Zhong ZC, Chen CS, Chen WJ, Hung C (2009) Biomechanical comparison between lumbar disc arthroplasty and fusion. Med Eng Phys 31(2):244–253
14. Chiyokura H (1988) Solid modelling with DESIGNBASE—theory and implementation. Addison-Wesley Publishing Company, Boston
15. Cowin SC (2001) Bone mechanics handbook, 2nd edn. CRC Press LLC, Boca Raton, FL
16. Cukovic S, Devedzic G, Ivanovic L, Zecevic Lukovic T, Subburaj K (2010) Development of 3D kinematic model of the spine for idiopathic scoliosis simulation. J Comput Aided Des Appl 7(1):153–161
17. DICOM Web site (2010) http://dicom.nema.org/
18. Eberlein R, Holzapfel GA, Fröhlich M (2004) Multi-segment FEA of the human lumbar spine including the heterogeneity of the annulus fibrosus. Comput Mech 34(2):147–163
19. Ern A, Guermond JL (2004) Theory and practice of finite elements. Springer, NJ
20. Ezquerro F, Simón A, Prado M, Pérez A (2004) Combination of finite element modeling and optimization for the study of lumbar spine biomechanics considering the 3D thorax–pelvis orientation. Med Eng Phys 26(1):11–22
21. Geomagic (2009) Geomagic studio 11: knowledge base—online support. http://support1.geomagic.com/ics/support/default.asp?deptID=5668&task=knowledge&questionID=883
22. Glossary of Spinal Terms (2009) http://www.spineuniverse.com/anatomy/glossary-spinal-terms. Accessed 20 Aug 2010
23. Gonugunta V, Krishnaney AA, Benzel EC (2005) Anterior cervical plating. Neurol India 53(4):424–432
24. Gonzalez RC, Woods RE (2008) Digital image processing, 3rd edn. Pearson Prentice Hall, NJ
25. Halldin P, Brolin K, Hedenstierna S (2008) Finite element model of the human neck. http://www.kth.se/sth/forskning/forskningsomraden/2.3593/projekt/2.3790/finit-elementmodellering-av-den-manskliga-nacken-1.14459?l=en_UK
26. Harms J (2010) Spinal disease information portal. http://www.harms-spinesurgery.com/
27. Henry RW (2007) Silicone plastination of biological tissue: cold-temperature technique. North carolina technique and products. J Int Soc Plastination 22:15–19
28. Hoffmann CM (1997) Solid modeling. In: Goodman JE, O'Rourke J (eds) CRC handbook on discrete and computational geometry. CRC Press, Boca Raton, FL, pp 863–880, http://www.cs.purdue.edu/homes/cmh/distribution/papers/SolidModel/56.pdf
29. ICB-Dent Web site (2005) Integration of clinical and biomechanical competencies into on-line course material for prosthetic dentistry and implant dentistry. http://www.feppd.org/ICB-Dent/index.htm
30. Iliescu A, Gavrilescu D (1976) Anatomia funcțională și biomecanică. Editura Sport-Turism Bucuresti
31. Kadoury S, Paragios N (2009) Surface/volume-based articulated 3D spine inference through markov random fields. In: Proceedings of the 12th international conference on medical image computing and computer-assisted intervention: Part II. Springer, Berlin, Heidelberg, pp 92–99
32. Klinder T, Ostermann J, Ehm M, Franz A, Kneser R, Lorenz C (2009) Automated model-based vertebra detection, identification, and segmentation in CT images. Med Image Anal 13(3):471–482
33. Kohnke P (2001) Ansys Inc. Theory manual, 12th edn. SAS IP, Inc. U.S.A., pp 27–49, 1040–1049

34. Kurtz SM, Edidin AA (2006) Spine technology handbook. Elsevier Academic Press, New York
35. LaCourse DE (1995) Handbook of solid modeling. McGraw-Hill, Inc., New York
36. Li H, Wang Z (2006) Intervertebral disc biomechanical analysis using the finite element modeling based on medical images. Comput Med Imaging Graph 30(6):363–370
37. Lodygowski T, Witold K, Wierszycky M (2005) Three-dimensional nonlinear finite element model of the human lumbar spine segment. Acta Bioeng Biomech 7(2):17–28
38. Materialise (2009) Mimics 10.01 support files. http://www.materialise.com/BiomedicalRnD/training
39. Owen SJ, Canann SA, Saigal S (1997) Pyramid elements for maintaining tetrahedra to hexahedra conformability, AMD-Vol.220. In: Proceedings special session on trends in unstructured mesh generation: 1997 Joint ASME/ASCE/SES Summer Meeting, Northwestern University, Evanston, IL, 29 June–2 July 1997, pp 123–129
40. Panjabi MM (1998) Cervical spine models for biomechanical research. Spine 23(24):2684–2700
41. Papilian V (2006) Anatomia omului, vol. I: aparatul locomotor. Editura ALL bucuresti
42. Pianykh OS (2008) Digital imaging and communications in medicine (DICOM): a practical introduction and survival guide. Springer, Berlin, Heidelberg
43. Pitzen T, Kettler A, Drumm J, Nabhan A, Steudel WI, Claes L, Wilke HJ (2007) Cervical spine disc prosthesis: radiographic, biomechanical and morphological post mortal findings 12 weeks after implantation. A retrieval example. Eur Spine J 16(7):1015–1020
44. Requicha A (1996) Geometric modeling: a first course. http://www-pal.usc.edu/~requicha/book.html
45. Rhinoceros (2007) NURBS modeling for Windows. Modeling tools for designers. http://www.rhino3d.com/nurbs.htm
46. Sora MC (2007) Exposy plastination of biological tissue: E12 Ultra-thin technique. J Int Soc Plastination 22:40–45
47. Standring S (2005) Gray's anatomy: the anatomical basis of clinical practice, 39th edn. Elsevier, Churchill-Livingstone
48. Starly B, Darling A, Gomez C, Nam J, Sun W, Shokoufandeh A, Regli W (2004) Image based bio-cad modeling and its applications to biomedical and tissue engineering. In: Proceedings of the ninth ACM symposium on solid modeling and applications, Eurographics association, Switzerland, pp 273–278
49. Starly B, Fang Z, Sun W, Shokoufandeh A, Regli W (2005) Three-dimensional reconstruction for medical-CAD modeling. J Comput Aided Des Appl 2(1–4):431–438
50. Stoia DI (2008) Modelarea, dezvoltarea si testarea implanturilor pentru coloana vertebrala. Editura politehnica timisoara
51. Stoia DI, Toth-Tascau M (2010) Static equilibrium of the cervical spine. Annals of the Oradea University, Fascicle of Management and Technological Engineering, vol IX(XIX), pp 1163–1168
52. Stryker Web site (2010) Reflex hybrid anterior cervical plate system. http://www.stryker.com/en-us/products/Spine/Cervical/ReflexHybridACP/index.htm
53. Sui HJ, Henry RW (2007) Polyester plastination of biological tissue: Hoffen P45 technique. J Int Soc Plastination 22:78–81
54. Sun W, Lal P (2002) Recent development on computer aided tissue engineering—a review. Comput Methods Programs Biomed 67(2):85–103
55. Sun W, Darling A, Starly B, Nam J (2004) Computer-aided tissue engineering: overview, scope and challenges. Biotechnol Appl Biochem J 39(1):29–47
56. Sun W, Starly B, Nam J, Darling A (2005) Bio-CAD modeling and its applications in computer-aided tissue engineering. Comput Aided Des 37(11):1097–1114
57. Synthes Web site (2010) Products. http://us.synthes.com/Products/Spine/Cervical/
58. Taylor LD (1992) Computer aided design. Addison Wesley, England. Accessed 18 Feb 2010
59. Tejszerka D, Gzic M (1999) Biomechanical model of the human cervical spine. Acta Bioeng Biomech 1(2):39–48
60. Teo EC, Ng HW (2004) The biomechanical response of lower cervical spine under axial, flexion and extension loading using FE method. Int J Comput Appl Tech 21(1/2):8–15

61. Thacker BH, Nicolella DP, Kumaresan S, Yoganandan N, Pintar FA (2001) Probabilistic finite element analysis of the human lower cervical spine. J Math Modeling Sci Comput 13(1–2):12–21
62. Volker KH, Sonntag MD (2008) Discovery of the spine specialist: instrumentation of the cervical spine. Barrow Neurological Institute, Phoenix, Arizona
63. White AA, Panjabi MM (1990) Clinical biomechanics of the spine, 2nd edn. J. B. Lippincott, Philadelphia
64. Wills L, Cross JH (1996) Recent Trends and Open Issues in Reverse Engineering. Automated Software Engineering, 3, Kluwer Academic Publishers, Boston, pp 165–172
65. Wu JSS, Lin HC, Chen JH (2009) Geometrical nonlinear analysis of the spinal motion segments by poroelastic finite element method. In: Proceedings of computer science and information engineering, 2009 WRI World Congress, vol 5, Los Angles, 31 Mar–2 Apr 2009, pp 357–361
66. Yoganandan N, Kumaresan SC, Voo L, Pintar FA, Larson SJ (1996) Finite element modeling of the C4–C6 cervical spine unit. Med Eng Phys 18(7):569–574
67. Yognandan N, Kumaresan S, Voo L, Pintar F (1996) Finite element applications in human cervical spine modeling. Spine 21(15):1824–1834
68. Zhang QH, Teo EC (2008) Finite element application in implant research for treatment of lumbar degenerative disc disease. Med Eng Phys 30(10):1246–1256
69. Zimmer Web site Spine (2009) SC-AcuFix® anterior cervical plates. http://www.zimmer-spine.eu/z/ctl/op/global/action/1/id/10351/template/MP/navid/10202

Chapter 8
Carbon Nanotubes in Acrylic Bone Cement

Nicholas Dunne, Ross Ormsby, and Christina A. Mitchell

8.1 The Hip Joint

The hip joint is a synovial ball and socket joint allowing for rotation about three per-pendicular axes. It is constructed of the femoral head and the acetabulum of the pelvic bone. The femoral head and acetabulum are covered by cartilage. In a healthy hip joint the cartilage acts as a protective cushion to allow smooth movement of the joint, thus reducing friction and absorbing shock, to an extent. The presence of the synovial membrane secretes synovial fluid into the joint in order to nourish and lubricate the articulating cartilage [57]. The hip joint is responsible for the transfer of weight from the leg to the body, and as such, can be under enormous mechanical stresses.

8.1.1 Potential Problems with the Hip Joint and Primary Joint Replacement

Problems with the hip joint can arise due to cartilage damage within the joint caused by disease, trauma or congenital conditions. This can lead to the surrounding tissues becoming inflamed, causing considerable pain. If partial damage of the joint has occurred, it may be possible to repair or replace just the damaged areas; if the entire joint is damaged, however, a total joint replacement (TJR) may be necessary to

N. Dunne (✉) • R. Ormsby
School of Mechanical & Aerospace Engineering, Queen's University of Belfast,
Ashby Building, Stranmillis Road, Belfast, BT9 5AH, UK
e-mail: n.dunne@qub.ac.uk; rormsby01@qub.ac.uk

C.A. Mitchell
School of Medicine, Dentistry and Biomedical Sciences, Queen's University of Belfast,
Grosvenor Rd, Belfast, BT12 6BP, UK
e-mail: c.mitchell@qub.ac.uk

I. Antoniac (ed.), *Biologically Responsive Biomaterials for Tissue Engineering*,
Springer Series in Biomaterials Science and Engineering 1,
DOI 10.1007/978-1-4614-4328-5_8, © Her Majesty the Queen in Right of United Kingdom 2012

relieve pain and to maintain function of the joint [73]. When replacing a total joint, the diseased or damaged parts are removed and artificial parts, i.e. prostheses or implants, are fitted. TJR may be performed on a variety of joints, including hip, knee, ankle, shoulder, elbow, fingers and wrist. However, hip replacements are by far the most common, as reported, for example, in Norway between 1987 and 2004 [28].

During TJRs, the most commonly used method of implant fixation is with a load transferring grout-like material, typically an acrylic based bone cement. The major advantage of these cemented joint replacements is the reduced operation recovery time: once polymerised the cement is capable of bearing load and offers immediate stability. However, if the cement mantle becomes loose, the surrounding bone may resorb and ultimate failure of the implant may occur. Uncemented implants were introduced to overcome these shortcomings and others, such as cement wear particles, residual monomer and highly exothermic polymerisation reaction causing cellular necrosis of the surrounding bone. Uncemented implants typically use a roughened porous surface to promote bone growth around the prosthesis [39].

However, the bone cavity produced during the operation needs to be precisely shaped to ensure that the implant is held initially in place through an interlocking mechanical fixation between the implant and the bone. It is also essential that the surrounding bone is healthy to enable this technique to be successful. In addition, the recovery time is long as the bone is required to regenerate.

A combination of cemented and uncemented implants is also employed and often termed a "hybrid". More recently, resurfacing arthroplasty has been introduced, where less of the bone is removed compared with conventional TJR. Resurfacing procedures not only require the removal of less bone, but also cause fewer complications during revision surgeries because the femoral canal remains intact [2]. On average, the number of primary arthroplasties in developed nations is increasing each year [28, 43, 65].

Figure 8.1 demonstrates the proportion of cemented, uncemented and hybrid replacements. The total number of TJRs performed in England and Wales is significantly greater than Sweden and Norway. In 2004, for example, 48,987 THRs were recorded in England and Wales, compared to just 13,366 in Sweden and 7,061 in Norway. However, it should also be noted that the population of England and Wales (approximately 55 m) is significantly greater than Sweden and Norway (approximately 13 m). This equates to approximately 1 in every 1,100 people in England and Wales, and 1 in every 650 people in Sweden and Norway.

8.2   Acrylic Bone Cement

8.2.1   Composition

Acrylic bone cement is primarily composed of poly methylmethacrylate (PMMA). Most commercial acrylic bone cements comprise a two part self-curing acrylic polymer, usually formulated at a 2:1 powder-to-liquid ratio. These components are

Fig. 8.1 Number and
fixation type of primary
THRs preformed in Sweden
from 1979 to 2007 [43]

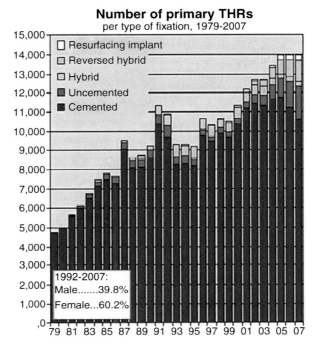

Number of primary THRs
per type of fixation, 1979-2007

mixed immediately prior to implantation during surgery and delivered directly to the implant site. The compositions of the main commercial bone cements are summarised in Table 8.1, showing variations in chemical composition.

Other cements may also contain antibiotics (e.g. gentamicin sulphate [34, 52]) in order to improve the body's response to the implant, reducing the risk of subsequent infection and implant rejection.

8.2.2   Polymerisation Reaction

PMMA is an amorphous polymer, which is plasticised on the addition of the monomer methyl methacrylate (MMA). When bone cement is mixed two processes occur. Firstly the monomer is absorbed by the PMMA beads and secondly, a free radical polymerisation reaction occurs [47]. This reaction is shown schematically in Fig. 8.2.

During this reaction the Dimethyl P Toluidine (DmpT) causes the benzoyl peroxide (BPO) to decompose leaving a benzoyl radical, and a benzoyl anion (Fig. 8.2a). These benzoyl radicals then initiate the polymerisation of the MMA by combining and forming an active centre (Fig. 8.2b). These active centres then combine with multiple molecules to form a polymer chain (Fig. 8.2c). This reaction forms a viscous fluid allowing the polymerising cement to be moulded as required. This is the stage when the orthopaedic surgeon would inject the bone cement into the prepared

Table 8.1 Compositions of six commercial formulations of bone cement [50]. The compositions are given in percent (w/w) except where stated otherwise

Constituent/bone cement type	CMW-1	CMW-3	Palacos R	Simplex P	Zimmer LVC
Powder components					
Benzoyl peroxide (BPO)	2.60	2.20	0.5–1.6	1.19	0.75
Barium sulphate (BaSO$_4$)	9.10	10.00	–	10.00	10.00
Zirconium dioxide (ZrO$_2$)	–	–	14.85	–	–
Chlorophyll	–	–	200 ppm	–	–
PMMA	88.30	87.80	–	16.55	89.25
PMMA–methacrylic acid (P(MMA/MA))	–	–	83.55–84.65	–	–
PMMA–styrene copolymers P(MMA/ST)	–	–	–	82.26	–
Liquid components					
N,N dimethyl P toluidine (N,N-DmpT)	0.40	0.99	2.13	2.48	2.75
Hydroquinone	15–20 ppm	15–20 ppm	64 ppm	75 ppm	75 ppm
Mehtylmethacrylate (MMA)	98.66	98.07	97.87	97.51	97.25
Ethanol	0.92	0.92	–	–	–
Ascorbic acid	0.02	0.02	–	–	–
Chlorophyll	–	–	267 ppm	–	–
Gentamicin sulphate	–	–	–	–	–

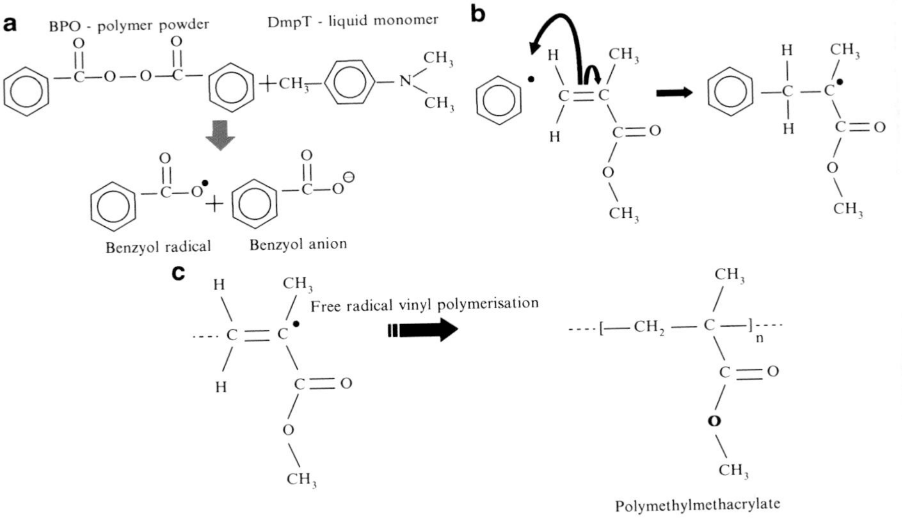

Fig. 8.2 (**a**) Schematic diagram showing the decomposition of BPO leaving a benzoyl radical, and a benzoyl anion; (**b**) how these benzoyl radicals initiate polymerisation of MMA; (**c**) formation of a polymer chain

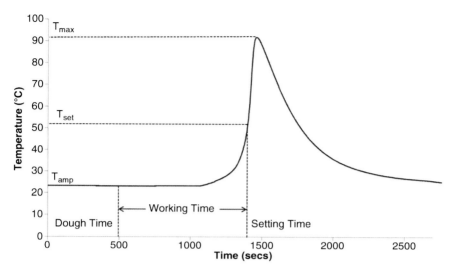

Fig. 8.3 A typical curing curve for acrylic bone cement where T_{max} is the maximum temperature reached, T_{set} is the setting temperature and T_{amb} is the ambient temperature

bone canal prior to implanting the stem. As the monomer begins to polymerise, the cement hardens around the orthopaedic implant, holding it in place. This reaction is highly exothermic; an example of a temperature plot of bone cement during polymerisation is shown in Fig. 8.3.

The heat energy produced during polymerisation is 57 kJ per mole MMA, resulting in temperatures, which can exceed 100 °C. These elevated temperatures can cause cellular bone necrosis and contribute ultimately to aseptic loosening [20, 47, 89]. It should be noted though that the polymerisation temperatures experienced in vivo are much lower (between 40 and 47 °C) at the bone interface [93]. This is due to reduced thickness of bone cement mantle, the presence of blood circulation and the dissipation of heat through the implant and surrounding tissue [47]. It has been shown that volumetric shrinkage can occur due to thermal contraction on cooling and the changing density as polymerisation progresses [30, 47]. Gilbert et al. [30] reported that volumetric shrinkage as a result of density variation, due to the exothermic polymerisation, was between 5.1 and 6.5 %, depending on mixing method employed and type of cement. Both shrinkage mechanisms have been identified as factors which influence the levels of residual stresses within the cement [30, 69].

As illustrated in Fig. 8.3, the time that has elapsed after initial mixing when the cement takes a homogeneous dough-like state is known as the "*dough time*". This point may be identified with temperature or average molecular weight of the polymer. However, as specified in the British Standard BS 7253 (ISO 5833:2002) [11], it is the point at which the cement will no longer stick to powderless surgical gloves (typically 2–3 min after initial mixing). The time from the end of dough time until the cement can no longer be manipulated is defined as the working time. During an operation this is the time during which the surgeon must insert the stem and adjust its position. Finally, the setting time is the time from the onset of mixing until the surface temperature reaches one-half of the maximum temperature, as described in ISO 5833:2002.

8.2.3   Thermal and Chemical Necrosis

As mentioned previously, in vivo temperatures during the exothermic polymerisation of bone cement can potentially cause thermal necrosis (tissue death) of the bone and impaired local blood circulation, which can lead to early failure through aseptic loosening of the implant [37]. It has been reported that for epithelial cell death to occur, an exposure time of 1 s is required for temperatures above 70 °C, 30 s for temperatures greater than 55 °C and approximately 5 h for temperatures greater than 45 °C [90]. Collagen protein molecules are denatured at 45 °C, and experience irreversible damage at 60 °C if held at these temperatures for an hour. It has also been reported that thermal necrosis occurs in bone tissue when exposure is greater than 1 min for temperatures above 50 °C and denaturation of sensory nerves occurs for temperatures above 45 °C if exposure exceeds 30 min. The amount of heat generated during polymerisation is dependent on the amount of reacting monomer; however the maximum temperature reached is also dependent upon the rate of heat dissipation. In vitro testing completed by Stanczyk and van Rietbergen [89] suggested that the tips of bone trabeculae protruding into setting cement may experience temperatures in excess of 70 °C. In the 1960s, upon the first use of bone cement in TJR surgery, Charnley believed that while temperatures of ~100 °C could be reached during polymerisation, in the presence of a metallic prosthesis, which would act as a heat sink, there was a reduction in the peak temperature experienced in vivo [13]. Since then, numerical simulations and in vitro studies of thermal necrosis and peak exotherms in TJR have helped to establish two methods which may assist the reduction of thermal necrosis: (a) the use of thin cement mantle layers and (b) pre-cooling of the bone surface [14, 27].

An additional potential adverse consequence of using standard acrylic bone cements is the leaching of residual liquid monomer into the surrounding tissue, which may cause inflammation, chemical necrosis and even death. Average levels of residual monomer can be as high as 5 %; however local concentrations may be as high as 15 %, increasing the likelihood of chemical necrosis [89]. Vacuum mixing of acrylic bone cement has been associated with reduced levels of residual monomer as mixing bone cement at reduced pressures increases monomer polymerisation [20].

8.2.4   Mechanical Properties

The main role of bone cement is to transfer load between bone and the orthopaedic implant. Several studies have shown that the composition of acrylic bone cement significantly influences the mechanical properties of the cement [33, 51]. It is during the polymerisation process that numerous cement properties, for example viscosity, setting time, maximum cure temperature, volumetric shrinkage, etc., can be determined. These material characteristics may influence the performance of a cemented TJR. It has also been shown that the variability in the mechanical static and dynamic properties of commercial bone cements is significant, with greater relative differences reported in fatigue properties [33, 50]. Harper and Bonfield [33] found that there was some correlation between the static and fatigue strengths; however the ranking of the different cements tested did not match exactly. Mechanical properties are known to be affected by cement composition, size and morphology of the PMMA beads, molecular weight, cement mixing technique and powder–liquid ratio [33, 50]. The variation in tensile strength, for example, is reported to vary between 24 and 49 MPa for five different commercial bone cement formulations, depending on the mixing technique, specimen age and test conditions [50].

8.2.5 Biocompatibility of Bone Cement "In Vivo"

Any biomaterial must exhibit some level of biocompatibility. An orthopaedic implant has the potential to alter both the mechanical and chemical environment locally (within the immediate surroundings) and systemically (throughout the whole body) [73]. The implantation of any material into the human body poses the risk of an immune response, involving inflammation of the local region, and potential rejection of the implanted material [34]. Traditionally PMMA has been accepted as a biocompatible or a bioinert material [5]. The potential for an immune response associated with the use of bone cement stems from the presence of unreacted residual MMA monomer; however, this MMA is rapidly removed from the body via the lungs, minimising systemic and cardiovascular reactions [19]. Osteolysis refers to an active resorption of bone due to disease, infection or inadequate blood supply [57]. Osteolysis is often associated with aseptic loosening, which has been shown to be the most common cause for revision surgery [43, 44, 82]. It has been reported that bone cement can lead to the onset of peri-prosthetic osteolysis (i.e. bone resorption around the implanted construct) in the form of small fragments [19, 32]. These small fragments may be generated as wear particles between articulating surfaces and, in excess, these particles are deposited in the surrounding tissue causing an inflammatory response. Additionally, wear particles may be generated during fatigue crack propagation of the cement mantle, and migration of these particles into the surrounding tissue is then possible [94].

8.2.6 Effect of Residual Stresses and Cement Shrinkage

It is well accepted that residual stresses are generated within the cement mantle following polymerisation and have a direct influence on the stress distribution at the cement–prosthesis interface. Knowledge and understanding of these processes may allow a more accurate prediction of load transfer and in-service conditions in the cement mantle [66, 76]. Stresses due to shrinkage exist in cement surrounding the stem in both the longitudinal and the hoop direction [76]. These stresses will be at their greatest immediately post-operatively, with a reduction occurring with time as stress relaxation and creep occur [66, 91, 95]. This may create more favourable stress distributions at the cement–bone and cement–prosthesis interfaces, as finite element modelling of bonded and un-bonded stems predicts that an increase in compressive stresses at these regions may occur [95]. The presence of pores at either interface has been attributed to volumetric shrinkage of the cement during polymerisation. Some polymerisation shrinkage will occur while the cement is still viscous and hence can be accommodated by flow [69]. However, as polymerisation progresses, this flow reduces shrinkage and cracks can initiate in high-stress areas as a mechanism of stress relief. While residual stresses do exist in fully polymerised bone cement, the additional presence of porosity, high stress concentrations or excessive heat generated during polymerisation may still be required for large cracks to initiate, as residual stresses alone may not be enough to generate cracks [49]. The ultimate tensile strength of various bone cements ranges between 24 and 49 MPa [50], whereas residual stresses of between 2.5 MPa [66] and 12.6 MPa [69] have been reported. The direction in which the cement will shrink is of great significance. Orr et al. [69] reported that this has a direct relation to the levels of micro-cracking that may occur as a result of shrinkage stress, although the use of acoustic emission has provided evidence for the shrinkage of the cement onto the femoral implants [76]. Furthermore, shrinkage is known to be affected by the volume fraction of monomer content [30].

8.2.7 The Role of Porosity

Lewis [50] identified four main reasons why porosity occurs in bone cements:

1. The entrapment of air between the polymer powder and monomer liquid as the powder is wetted by the monomer upon mixing
2. Evaporation of the liquid monomer during polymerisation
3. Entrapment of air during mixing
4. Entrapment of air upon transfer of the dough into the cement gun (depending on mixing method)

Reduced porosity allows for improved compressive, flexural and fatigue properties of acrylic bone cement [20, 21, 50, 62]; therefore the level of porosity (both macro- and micropores) should be minimised. Pores of diameter ≥1 mm are deemed macropores, and are generally introduced during the mixing process when air is trapped within the cement mixture. These macro-pores are often cited as being the cause of low fatigue life for test specimens as crack initiation is often associated with a single pore. Micropores have diameters ≤1 mm, and may be established due to the evaporation of the liquid monomer during the polymerisation process and/or entrapment

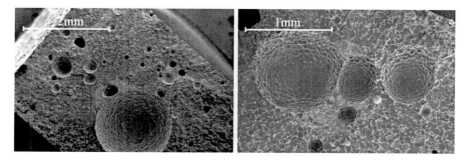

Fig. 8.4 Scanning electron micrograph of a hand-mixed fracture surface for a commercial bone cement showing a large pore with a large number of small pores [62]

Fig. 8.5 Scanning electron micrograph of a vacuum-mixed fracture surface of a commercial bone cement, with a larger pore size compared to hand-mixed cement [62]

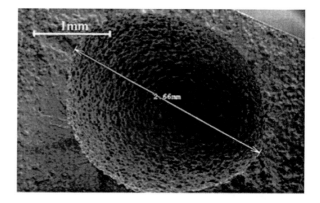

of air during mixing [21]. It is often observed that multiple smaller pores (≤ 1 mm) in close proximity are more detrimental than one larger pore (≥ 2 mm); this is often a result of the type of mixing method employed [62]. A multiple pore arrangement is typically observed for hand-mixed specimens (Fig. 8.4) where the combined interaction of the pores produces a stress concentration large enough to cause fatigue crack initiation. In contrast, vacuum mixing (Fig. 8.5) usually generates a smaller distribution of pores.

There is evidence to suggest that hand-mixed cement reduces the level of shrinkage that cement experiences due to the high level of porosity introduced during mixing [21], as only the cement shrinks during polymerisation and not the voids [47].

However, porosity may be beneficial to reduce residual stresses prior to loading; this benefit could be outweighed by the adverse effects observed for fatigue crack initiation and propagation. When the propagating crack tip reaches a pore, failure is considered to occur instantly across the void area, effectively causing the crack propagation rate to increase. Conversely though, there are studies that suggest that pores act as crack "blunters", thereby increasing the fatigue life of the bone cement [94], although crack acceleration into the void must also occur due to the local stress concentration at such a defect.

8.3   Reinforcement of Bone Cement

8.3.1   Developments in Acrylic Bone Cement

The intrinsic mechanical properties of acrylic bone cement (such as strength, fracture toughness and fatigue crack propagation resistance) in addition to the presence of extrinsic factors such as porosity, agglomerates of radiopaque agents and other such stress concentrations may limit its long-term survival [52]. Within the current literature there have been many attempts to improve the fatigue performance of bone cement reported. Most studies have tried to control the extrinsic factors, in particular porosity [52, 62, 64], by means of vacuum mixing or centrifugation. However, this does not address the underlying intrinsic factors which can be broadly categorised into two areas: (a) mechanical studies, focusing on improving mechanical performance and (b) biological studies where the focus may be on the effect of bioactive inclusions or the addition of antibiotics.

8.3.2   Mechanical Performance

A significant portion of the literature is directed towards discussing the potential to increase mechanical performance of acrylic bone cement via reinforcement with fibres or secondary phase particles: for example, carbon [70, 72, 75, 77, 98], polyethylene [63, 99], titanium [46, 94], hydroxyapatite [81], glass beads [86], glass flake [26], glass fibres [63] and steel [45]. Mechanical properties that have been reported to improve include compressive, tensile and bending strength, elastic modulus, fracture toughness and fatigue resistance, when compared to cement without reinforcement. In addition to the mechanical improvements provided by these fillers, further benefits have been identified with reduction in the peak temperature reached during polymerisation. High temperatures experienced in vivo can cause thermal necrosis of the bone cells surrounding the cement mantle, in addition to the coagulation of blood, which can potentially lead to aseptic loosening of the implant, and ultimately implant failure [50]. Reduced in situ polymerisation temperatures have been observed for, but not limited to, steel, carbon fibres and multiwalled carbon nanotube (MWCNT)-reinforced bone cement [45, 72, 77]. A number of researchers have investigated adding carbon fibre (CF) as a reinforcing agent using clinically applicable cement mixing techniques for both in vitro testing [70, 75, 77, 98] and in vivo applications [72]. Pilliar et al. [72] reported that the inclusion of randomly oriented CF (0.6 cm) improved fatigue performance, tensile strength, Young's modulus and impact resistance (i.e. indicative of toughness), compared to cement without reinforcement. The thermal properties were also observed for the two cement types; the dough time and the setting time were unaffected by the addition of CF, whilst the maximum polymerisation temperature was lowered for the cement with added CF (53 °C compared to 57 °C). Interestingly these research groups found that CF addition increased the viscosity of the cement above the required level by ASTM standards (ASTM F451—99a), meaning that use of these cements in a clinical setting would not be ideal. Fractographic analysis identified poor CF distribution, and evidence of poor CF-PMMA bonding, although fibre pullout was noted. This CF-reinforced cement was used in vivo with no detrimental mechanical or biological response observed after 18 months. During the 1980s, problems associated with the high starting viscosity of the cement, and subsequent reduced levels of intrusion, were investigated in vitro following the development of low-viscosity cement. Robinson et al. [75] confirmed that CF-reinforcement of a commercial cement increased fracture toughness of both regular and low-viscosity cements. However, the low-viscosity cements (both reinforced and conventional) displayed a reduction in fracture toughness when compared to the equivalent regular viscosity cement. For the reinforced cement, Wright and Robinson [98] reported a decreased crack growth rate versus the unreinforced cement. Saha and Pal [77] investigated the effect of mechanically or hand-mixed CF-reinforced bone cements, and found that mechanical mixing provided superior performance. They attributed this to the improved CF dispersion throughout the cement matrix.

8.3.3   Biological Performance

Bone cement is a biologically inert component of the implant construct. However, conventional acrylic bone cement does not normally promote bone ingrowth. Several studies however have attempted to improve the biological performance of bone cement. These have included the incorporation of bioactive agents, such as hydroxyapatite-based powders, glass ceramic particles or glass beads [48, 61, 86]. Each of these additives has been reported to enhance the biocompatibility of bone cement, thus reducing the formation of fibrous tissue at the bone–bone cement interface. Mousa et al. [61] used apatite-wollastonite glass ceramic particles to reduce the amount of monomer required for polymerisation, which lead to a reduction in the peak exotherm, and thermally induced bone necrosis, as well as decrease the levels of cement shrinkage. Similar results have been reported for cements containing glass beads, which have also been shown to improve bioactivity (in terms of osteoconductivity) compared with hydroxyapatite powder [86]. Additionally, it has been reported that many of these bioactive cement composites exhibited no detrimental influence on mechanical performance, and for some, reinforcements and improvement were observed [61, 86]. Concerns regarding biological performance include the use of antibiotics, which are integrated to reduce the risk of infection and associated revision [47]. Many antibiotic-loaded cements are currently commercially available and, those containing gentamicin sulphate, are believed not to cause any adverse affect on the fatigue performance [6].

8.3.4 Polymer Matrix Composites

The mechanical success of any polymer composite is governed by the successful transfer of load between the matrix and the reinforcement. This transfer of load is dependent upon the volume fraction, dispersion, orientation of the reinforcing phase, host matrix–reinforcement interface and individual mechanical properties of the phases that are present [29, 38, 99]. Within fibre-reinforced composites four main microstructural regions exist: (1) the matrix, (2) the fibre, (3) the interface and, in some composites, (4) the interphase. An interphase may be present if a mechanical or chemical interaction takes place between the polymer matrix and the reinforcing phase. Examples include adsorption of the polymer onto the surface of the reinforcing agent in particulate-reinforced polymers, inter-diffusion of the components during blending and chemical reactions at the polymer/fibre interface [74]. Fibres aligned in the direction of applied load are particularly effective at reinforcing composites. Corresponding mechanical properties which effect failure performance may be identified, with a complex interaction between individual phase properties, the interfacial strength between the host polymer and the reinforcing fibres and the composite microstructure. Polymers are the most common form of composite matrix and are often reinforced with a low fraction of fillers such as glass, CF or aramid. This results in composites of high specific strength and modulus as the low levels of additives allow a more homogeneous dispersion [12]. Of these three reinforcements, CF–polymer composites often exhibit the best resistance to fatigue failure due to superior mechanical properties as well as the higher thermal conductivity of the CF, which assists in the dissipation of heat during cyclic loading [79]. In compression, the mechanical performance of fibre-reinforced composites is dependent on the interaction between the host polymer matrix and the fibre. For optimum reinforcement, the matrix would provide lateral stabilisation to the fibre preventing subsequent buckling. Alternatively, the tensile behaviour is governed by the tensile strength of the fibre additive [38]. Fatigue failure in polymer composites is commonly characterised by a gradual reduction in stiffness [80]. Without reinforcement, fatigue failure typically occurs perpendicular to the applied load. In contrast, the presence of fibres generally results in a diffuse damage zone due to the combination of a number of subcritical failure modes and crack shielding mechanisms. In general, crack propagation through fibre-reinforced polymers may be considered as a multifaceted interaction between the polymer matrix, the fibre reinforcement and the associated interface/interphase regions. A combination of mechanisms may occur and subsequently, fibre inclusions may impede crack growth by three main mechanisms [54, 78]:

1. Debonding of interface/interphase between fibre and matrix—as a crack approaches, failure of the interface occurs serving to blunt the crack tip and reduce crack propagation.
2. Crack bridging—transferring load across a given matrix crack, reducing the crack.
3. Fibre pullout, subsequent to crack bridging, may also absorb energy due to matrix deformation and/or interface friction.

8.3.5   *Carbon Nanotubes*

CNT powders can be classified as single-walled nanotube (SWCNT) or multiwalled nanotube (MWCNT) powders. SWCNT powder consists of a single graphene sheet rolled up as a seam-free tube. They can be thought of as a linearly extended fullerene [1, 7, 40, 41]. SWCNTs usually exist as agglomerations due to the van der Waals forces between each tube, with diameters on average between 0.7 and 2 nm, whilst their lengths are often 5–30 μm [1, 16]. MWCNT powders can be thought of as an array of SWCNT that are concentrically arranged inside one another with an internal diameter as small as 2.2 nm. The distance between the individual SWCNT that constitute MWCNT (or the graphite inter-layer separation) is typically 0.34 nm [1, 40, 41] (Fig. 8.6). Iijima [40] used electron diffraction to establish that the crystal axis of the graphene tubes consisted of carbon-atom hexagons arranged in a helical manner about the tube axis (Fig. 8.7). The CNT ends are closed off by the presence of pentagonal carbon rings near the tip regions, whilst deformations and imperfections of the cylinder occur as pentagons or heptagons within the main structure of the tube [1].

SWCNTs are known to be stiff and exceptionally strong (high Young's modulus and tensile strength). Furthermore, SWCNTs can stretch beyond 20 % of their original length and bend over double without kinking [7, 16]. However, the mechanical properties of individual CNT, whether of SWCNT or MWCNT arrangement, are the subject of much research with a significant variation in recorded properties existing. While direct measurement of mechanical properties for individual CNT remains challenging and high degrees of experimental uncertainty exist, it is clear that the use of CNT in various matrices can greatly enhance mechanical properties.

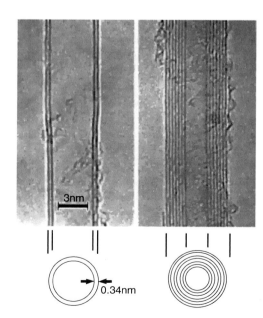

Fig. 8.6 Transmission electron micrograph (TEM) of different MWCNT arrangements [40]

Fig. 8.7 Helical arrangement
of carbon atom hexagons that
make up a graphene sheet in
a MWCNT [40]

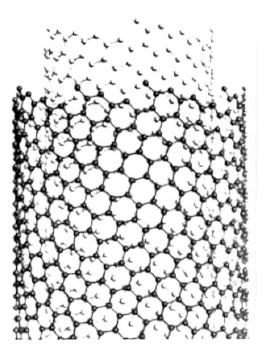

Wong et al. [97] provided an insight into the potential applications of CNT. Atomic force microscopy (AFM) was employed to determine the elasticity, strength and toughness of individual silicon carbide (SiC) nanorods and MWCNT that were attached to molybdenum disulphide surfaces. The average bending strength of the MWCNT was 14.2 ± 8.0 GPa, with the maximum bending strength being substantially smaller than that of the SiC nanorods at 53.4 GPa. In contrast, whilst both nanostructures exhibited high values for Young's modulus (highlighting their suitability as reinforcing agents in ceramic, metal and polymer matrix composites) the Young's modulus for the MWCNT was almost double that of the SiC nanorods (1.28 ± 0.59 TPa and ~600 GPa, respectively). The Young's modulus value for the in-plane modulus of highly oriented pyrolytic graphite was recorded at 1.06 TPa [9] and is believed to be the largest known for a bulk material. Wong et al. [97] concluded that while the stiffer MWCNTs had a lower ultimate strength, the elastic buckling displayed by the MWCNTs (i.e. the energy storing capabilities of CNT before failure) showed them to be the "tougher" nanostructure. More recently, Demczyk et al. [18] investigated the direct failure of individual MWCNT under tension using TEM. While the mode of failure, either ductile or brittle, could not be determined, results confirmed a tendency to fail via a mode now known as "telescopic failure", with initial failure observed in the outermost walls followed by a "sword in sheath effect" of the inner cylinders. These TEM observations also confirmed the elastic capabilities of MWCNTs during deformation in bending, even after being highly distorted.

It is clear that the incorporation of CNT offers significant potential to improve the properties for many existing materials; the challenge remains for the superior properties exhibited by CNT individually to be successfully applied and optimised in practical applications such as nanocomposites. Not only do the properties of the CNT themselves vary due to impurities during processing [7], but on addition of CNT to a matrix material, the problem becomes multifaceted: CNT dimensions, dispersion, alignment, concentration, CNT–matrix interface/adhesion and choice of matrix are some of the many issues that govern the final properties of the composite material. At present, production of high-purity SWCNT still remains costly and of a low yield: purification reduces yield further, adds to cost and damages the structure [3]. Low manufacturing costs and high yields of aligned MWCNTs are now possible however, making them the preferred choice in bulk composite material development.

8.3.6   CNT-Reinforced Biomaterials

Nanotechnology in biomaterials is not a new idea [36]. Nanomaterials have been used as implant coatings, bulk materials, in drug delivery, actuators, diagnostic tools and devices [87]. When biomaterials incorporating nanomaterials are studied, much of the emphasis is on the interaction between the biological tissue and the biomaterial at a molecular level. Using the interface between bone, and a metallic implant as an example, a positive biological interaction is essential if a good fixation is to be obtained. Chun et al. [15] examined this interaction by coating titanium (Ti) substrates with helical rosette self-assembled organic nanotubes (HRN). HRNs display chemical and structural similarities to various constituents of bone (Fig. 8.8). Chun et al. [15] found that the HRN-coated titanium displayed enhanced interaction with the naturally occurring nanostructures' constituents such as collagen fibres and

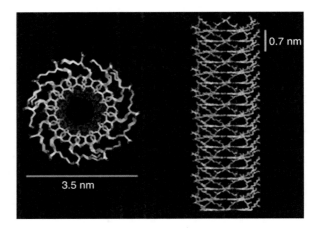

Fig. 8.8 Diagram of helical rosette nanotube (HRN) [15]

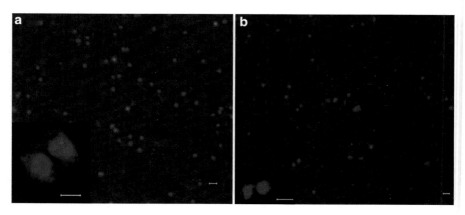

Fig. 8.9 Fluorescently stained cells on Ti substrates. (**a**) HRN-coated Ti. (**b**) Uncoated Ti (magnification: 20×; inset magnification 200×). Bars = 60 μm, *inset* bars = 50 and 100 μm for (**a**) and (**b**), respectively [15]

hydroxyapatite; this was measured as a function of the cell adhesion of human fetal osteoblast (hFoB) cells (Fig. 8.9).

Webster et al. [96] incorporated carbon nanofibres (CNF) into polycarbonate-urethane for neural or orthopaedic prosthetic devices. They reported that this material had the potential to increase neural and osteoblast functions, as cell attachment increased with CNF loading. Additionally they stated that the functions of cells that contributed to glial scar-tissue formation for neural prostheses (astrocytes) and fibrous-tissue encapsulation for bone implants (fibroblasts) decreased on the polycarbonate-urethane composites containing increasing amounts of CNF. In this manner, this study provided the first evidence that CNF formulations may interact with neural and bone cells, which is important for the design of successful neural probes and orthopaedic implants.

Furthermore, Webster et al. [96] summarised that using nanotechnology in biological systems may be potentially feasible as biological systems are governed by molecular behaviour at the nanoscale, and therefore the properties of which are accustomed to high levels of interaction at this nanoscale. This study by Webster et al. highlighted the potential for CNT to be used in PMMA bone cement to encourage cell growth at the bone–cement interface with the aim of reducing aseptic loosening by enhancing the mechanical interlock in the cancellous bone. As mentioned previously, CNTs exhibit many unique mechanical, thermal and electrical properties. However, their potential use for biomaterials or biomedical applications is almost wholly dependent on their biocompatibility. Cui et al. [17] investigated the effect of SWCNT on human HEK293 cells (human embryo kidney cells). Results showed that SWCNT could inhibit HEK293 cell proliferation, inducing cell apoptosis (programmed cell death as controlled by the nuclei in normally functioning cells) and decreasing cellular adhesive ability in a dose- and time-dependent manner. Their results also showed that HEK293 cells initiated active responses such as secretion of small "isolation" proteins to isolate the cells attached to the SWCNT from the rest of the cell mass, a response that offers potential for medical chemistry and disease therapy. Synthetic bone scaffolds is an area where the biocompatibility of materials used, such as polymers or peptide fibres, is still an issue where possible rejection by the body is feasible. CNTs offer mechanical advantages over the polymers or peptide fibres currently used in bone scaffolds. Zhao et al. [100] investigated the use of chemically functionalised SWCNT as a scaffold material for the growth of artificial bone and they identified the potential for the self-assembly of hydroxyapatite on the SWCNT surface. They suggested that this was possibly due to the presence of negatively charged functional groups on the SWCNT that attracted the calcium cations present in the hydroxyapatite [100]. The group also proposed that it is the high tensile strength, high degree of flexibility and low density of CNTs that make these materials ideal for the production of bone. The diameters of SWCNT used in the study by Zhao et al. [100] are of similar order and magnitude to the triple-helix collagen fibres within bone, and as such can act as scaffolds for the nucleation and growth of hydroxyapatite.

The potential for CNT to be used within biomaterials and biomedical applications is by no means endless; however while many more applications could be discussed, the following papers offer further information on the use of nanotechnology for biomedical applications [8, 87, 96]. Investigations concerning the cytotoxic response of CNT-containing materials have reported encouraging results confirming their potential use in orthopaedic applications [88]. However, many questions remain unanswered and as yet, the understanding of the toxicity and biocompatibility of CNT-reinforced materials is not fully established. More recent studies completed by Marrs et al. [56] have shown that the addition of MWCNT to PMMA bone cement may significantly improve the mechanical and fatigue properties, with an optimal loading of 2 wt.% MWCNT [55]. Marrs et al. achieved a high level of homogeneous dispersion by high-speed shear mixing and vacuum hot-pressing. However, these methods are of limited use in TJR applications— both studies used non-clinically relevant methods to incorporate the MWCNTs into the cement. Marrs et al. dispersed the nanotubes through molten PMMA using stainless steel counterrotating rotors in a mixing chamber at 220 °C. The molten PMMA was then collected and allowed to cool under ambient conditions until solid, thus ensuring maximum dispersion of the MWCNTs in the cement matrix and a reduction in the incidence of entanglements. Following full polymerisation, the cement was crushed into pellets and then hot-pressed under vacuum into thin sheets [56]. Additionally, non-standardised test methods were used for specimen preparation.

Ormsby et al. [67] overcame the limitations of these studies by Marrs et al., by incorporating MWCNT (either unfunctionalised, or carboxyl functionalised) into PMMA bone cement via three clinically relevant techniques. Ormsby et al. [67] added the MWCNT to the liquid monomer phase via either magnetic stirring, or ultrasonic dispersion, or to the polymer powder phase via turbo blending. They reported that the addition of MWCNTs (0.1 wt.%) could significantly improve the static mechanical properties, in addition to significantly reducing the exotherm of the polymerisation reaction. Ormsby et al. [68] stated that the dispersion of the MWCNT is critical in improving the mechanical properties of the bone cement composites. This is thought to be a significant finding as mechanical failure of the bone cement mantle remains a major problem in joint replacement surgery [94]. As is the case with typical fibre-reinforced composites, mechanical failure of PMMA bone cement is thought to take place in three phases:

- Crack initiation due to an initial imperfection in material stability,
- Slow crack growth and/or
- Rapid propagation to fracture

MWCNT addition to the liquid monomer by magnetic stirring was shown to have an overall negative effect on the mechanical performance of the bone cement. This may be attributed to poor MWCNT dispersion within the liquid monomer, resulting in MWCNT agglomerations within the cement matrix. Such agglomerations acted as stress concentrations within the cement, providing a mechanism for premature failure of the cement when subjected to a loading regime. In contrast, dry blending of the MWCNT into the polymer powder or disintegrating the MWCNT in the liquid monomer using ultrasonic agitation increased the disentanglement of the nanotubes and allowed for a more homogeneous dispersion of the MWCNT in the resulting nanocomposite. Andrews and Weisenberger [4] reported that ultrasonic dispersion was an effective method for dispersion of MWCNT at low levels (<5 wt.%) of concentration [4]. However, Marrs et al. [56] stated that care is needed when dispersing MWCNT within a polymer matrix. Furthermore, they went on to detail the adverse effects of inadequately dispersed clumps of MWCNT, particularly at levels of loading greater than 5 wt.%.

The presence of well-dispersed MWCNT within the PMMA bone cement resulted in a percentage of the MWCNTs being orientated with their longitudinal axis perpendicular to the crack wave. Such MWCNTs were effective in bridging the initial crack and preventing crack propagation, with the potential, in theory for further enhancing the longevity of the cement mantle. These improvements reported by Ormsby et al. [67] are clinically beneficial for the use of reinforced PMMA bone cement in total joint arthroplasties due to a reduction in the rate of crack propagation. This improvement may be most important for improperly placed femoral implants with thinner cement mantle layers, which continues to be cited as a factor that may reduce joint replacement longevity [60].

Previous studies have reported reductions in the thermal properties of PMMA cement on addition of 5–15 wt.% steel fibres [45]. The importance of minimising the exothermic reaction is significant, as it may result in a permanent cessation of blood flow and bone tissue necrosis, which shows no sign of repair after 100 days [23, 25, 58, 59]. The cumulative thermal necrosis index (TNI) has been used previously to assess the level of irreparable damage bone cement caused by heat generation [20, 59]. If the TNI value exceeds one, then there is the possibility of thermal damage to the living tissue cells. The TNI was calculated for two temperatures,

>44 °C and >55 °C. These were purposefully chosen as the temperature threshold for impaired bone regeneration has been reported to be in the range of 44–47 °C [23, 59]. The incorporation of MWCNT to PMMA bone cement may reduce the incidence of polymerisation-induced hot spots resulting in thermal necrosis of the surrounding tissue adjacent to the cement mantle, which is believed to be observed radiographically [53]. Reducing the occurrence of such tissue damage may improve the mechanical integrity of the bone cement–bone interface, thereby promoting implant longevity.

Ormsby et al. [68] also recorded that the incorporation of MWCNTs assisted in the dissipation of the heat produced during the exothermic polymerisation reaction of PMMA bone cement. The addition of carboxyl functionalised MWCNT to the liquid monomer using magnetic stirring was successful in reducing the TNI values at >44 °C and >55 °C to levels below one. It is proposed that the MWCNT acted as a heat sink within the PMMA bone cement and therefore contributed to dissipation of the heat produced during the polymerisation reaction by increasing the thermal conductivity of the cement. Bonnet et al. [10] reported a 55 % increase in thermal conductivity of PMMA on addition of 7 vol.% of SWCNTs [10]. It is hypothesised that the reductions reported here are largely due to the heterogeneous nature of the MWCNTs within the PMMA cement, which significantly reduces the mechanical properties of the cement. The thermal properties of the MWCNT–PMMA bone cement composites can be discussed in terms of the successful dispersion of MWCNT within the cement microstructure. As for electrically or thermally conductive MWCNT composites, the MWCNTs must form a percolated network of overlapping or touching CNT to allow for the transport of electrons, phonons and heat energy [56]. Therefore, the cement with the poorest mechanical properties, i.e. cement that has been magnetically stirred, demonstrated the optimal thermal properties. This is due to poor dispersion of MWCNT and subsequent agglomerations. Conversely, the MWCNT–PMMA composite cement that provided the greatest improvements in mechanical properties (ultrasonic dispersion), attributed to the successful dispersion of MWCNT, actually increased the polymerisation exotherm.

Ormsby et al. [68] also investigated the influence of adding MWCNT of varying chemical functionalities at different loadings (0.1, 0.25, 0.5, 1.0 wt.%) to PMMA bone cement as a means of improving MWCNT dispersion and mechanical properties. Dunne et al. [22] characterised the rheological and handling properties of the same composite cements. The MWCNTs were either unfunctionalised or amine (NH_2) or carboxyl (COOH) functionalised. Adding MWCNT at loadings ≤0.25 wt.% improved the mechanical properties of the resultant nanocomposite cement. Adding carboxyl and amine functionalised MWCNT enhanced the dispersion of the MWCNT within the cement matrix and, as a result, increased the interaction between the carbon nanotubes and the cement, thereby improving the mechanical integrity of the resultant nanocomposite cement. Adding MWCNT at higher loadings (≥0.5 %wt.) provided a negative effect on the mechanical performance of the MWCNT–PMMA cement. This was attributed to poor dispersion of the MWCNT resulting in the formation of agglomerations within the cement matrix. These agglomerations acted as stress concentrations within the cement microstructure, therefore providing a mechanism for premature failure of the cement when subjected to load. In contrast, low loadings (≤ 0.25 wt.%) of MWCNT were more readily disentangled by the application of ultrasonic energy and homogenously dispersed in the resulting nanocomposite. The presence of well-dispersed MWCNT in PMMA cement was effective in bridging the initial crack and preventing crack propagation, thus further enhancing the mechanical integrity of the cement mantle. Dunne et al. [22] recorded that the increased loading of MWCNT extended the time at which polymer gelation occurred, as well as extended the time before the onset of polymerisation. They attribute this to a chemical interaction between the functional chemical groups on the MWCNT, in addition to the higher loadings of MWCNT preventing the polymer chains from forming [22].

Gojny et al. [31] reported that the addition of chemical functional groups to MWCNT can provide a negative charge to the MWCNT and thus reduce agglomeration. This in turn improves interaction between the nanotubes and the host polymer. The results of these studies by Ormsby et al. [67, 68] concurred with the findings of Gojny et al. [31]. The PMMA bone cement with unfunctionalised MWCNT exhibited the least significant improvements for all mechanical properties measured. This reduced improvement was attributed to poor MWCNT dispersion within the cement matrix, resulting in the occurrence of MWCNT agglomerations. The unfunctionalised MWCNT provided a degree of mechanical reinforcement at lower loadings (≤ 0.25 wt.%), largely due to the reduced tendency for MWCNT agglomerations to occur. In the research conducted by Ormsby et al. [67, 68], MWCNTs chemically functionalised using either carboxyl or amine groupings were chosen, as it has been reported that the mechanical properties of generic PMMA can be improved using MWCNT with these functional groups [71]. MWCNT functionalised with carboxyl groups provided the most significant improvements in all mechanical properties of the PMMA cement. It is proposed that these significant improvements are a result of homogeneous MWCNT dispersion within the PMMA matrix, aided by negatively charged carboxyl groups. This homogeneous dispersion, in tandem with interfacial interactions between the functionalised MWCNT and PMMA matrix, could provide improved mechanical properties within the resultant nanocomposite. MWCNT–PMMA composite cements incorporating amine functional groups also exhibited improved mechanical properties. These improvements were less significant (p-value ≤ 0.01) however, compared to the PMMA cement containing carboxyl functionalised MWCNT. It was postulated that this was due to the lower level of functional groups present on the amine functionalised MWCNT when compared with the carboxyl functionalised MWCNT (that is 0.5 % versus 4.0 %, functional groups, respectively). This lower concentration of amine functional groups may result in a more heterogeneous MWCNT dispersion within the cement matrix, therefore resulting in a less successful transfer of stress through the cement mantle.

8.4   Biocompatibility of CNTs

The Royal Society and the Royal Academy of Engineering, UK, published a report discussing the associated ethical, health and safety, and social implications of nano-technology [92]. With an increased interest in the application of nanotechnology, the UK Government later published its own report ("*Characterising the potential risks posed by engineered nanoparticles*", November 2005) and follow-up studies ("*First quarterly update on the Voluntary Reporting Scheme for engineered nano-scale materials*", December 2006). The full report addressed many issues concerning the potential use of nanotechnology and CNT. Concerns have been raised that the properties that promote the use of nanoparticles in certain applications may also have health implications, such as their high aspect ratios, surface reactivity and their ability to cross cell membranes [24, 42, 92]. The report highlighted that the main risks associated with CNTs stem from their high surface area to which a target organ may be exposed, in addition to the chemical reactivity of the surface, the physical dimensions of the nanoparticles and their solubility. Speculation surrounding the use of CNTs has equated their effect on health to that of asbestos (due to their similar size and shape). CNTs are therefore suspected as being potentially carcinogenic, and additionally, may cause inflammation or functional changes to proteins due to their large surface area. However, it has been argued that no new risks to health have been introduced as a result of the increasing use of nanoparticles as part of composite materials, and that most concerns derive from the possibility of detached or "free" nanoparticles and nanotubes from the matrix [92]. It is believed that, if airborne, the likelihood of CNTs existing as individual fibres is improbable as electrostatic forces cause the CNTs to agglomerate which reduces their ability to be inhaled into the deeper areas of the lungs. However, when investigating the inhalation of stable non-purified SWCNT aerosols in mice, Shvedova et al. [84] reported that the chain of pathological events was realised through an early inflammatory response and oxidative stress culminating in the development of multifocal granulomatous pneumonia and interstitial fibrosis (Fig. 8.10).

Smart et al. [88] reviewed the often conflicting findings pertaining to the cyto-toxicity and biocompatibility of CNT. They concluded that, as-received (i.e. untreated, or unfunctionalised) CNT exhibited some degree of toxicity (observed both in vitro and in vivo) with detrimental effects associated with the presence of transition metal ions, used as catalysts in the CNT production. Smart et al. [88] also reported that functionalised CNTs have yet to demonstrate toxicity effects. It was highlighted that the tendency for CNT to aggregate may have impacted the reported results, although quantification of this fact has yet to be investigated. With research into the use of CNTs, and nanotechnology ever increasing, the uncertainty regarding toxicity has been brought to public attention. As a result, it has been recommended that further research is necessary regarding the biological impacts of nanoparticles and nanotubes, including their exposure pathways within the body, and that methodologies for in situ monitoring should also be developed [92]. Recent studies have addressed the issue of CNT uptake by different cell types.

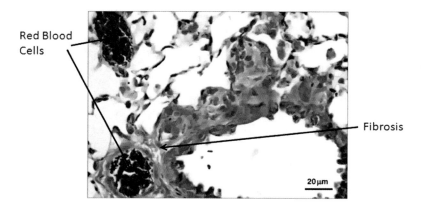

Fig. 8.10 Representative image of lung section from the SWCNT inhalation study depicting granuloma formation on day 28 post treatment. Fibrosis is indicated by blue staining in this Masson's Trichrome stained section of the lung [84]

While the results seem to be controversial, it is apparent that the presence or absence of specialised signals determined the recognition and subsequent interactions of CNT with cells. Overall, pristine CNTs carrying no recognisable signals were poorly taken up whereas CNTs modified chemically (e.g. oxidatively modified and functionalised) or by adsorbed macromolecules (e.g. proteins, lipids) were more readily recognised and engulfed by cells [85]. Several in vitro studies support the concept that pristine CNTs are not readily taken up by lung cells. Herzog et al. [35] reported no uptake of CNT in either A549 cells or BEAS-2B cells (a human bronchial epithelial cell line). Lastly, no evidence of CNT uptake was reported after electron microscopic evaluation of exposed RAW 264.7 cells (a mouse peritoneal macrophage cell line) [83]. In contrast, functionalisation of SWCNT with a phospholipid signal, phosphatidylserine, made CNTs recognisable in vitro by different phagocytic cells, including murine RAW264.7 macrophages, primary monocyte-derived human macrophages and dendritic cells, and microglia from rat brain (Fig. 8.10) [85].

8.5 Summary

The number of primary TJRs continues to increase each year and, even with the reported decrease in the proportion of cemented TJRs performed, PMMA bone cement is still required for the majority of TJR procedures. At present, with longer life expectancy and younger patient populations requiring TJR, an increase in cemented revisions seems inevitable. Aseptic loosening is continually cited as being the most common indication for revision. It is well established that for cemented implants a number of factors contribute to aseptic loosening, of which fatigue damage of the cement mantle has been observed in vivo. Fibre-reinforced materials with a high degree of fibre–matrix interaction have been shown to increase the fracture resistance of many polymer matrices: for example, CF-reinforced bone cement exhibits reduced fatigue crack propagation rates due to mechanisms such as crack bridging, telescopic failure and fibre pullout due to interface failure. Furthermore, it has been reported that fibre-reinforcement of PMMA bone cement leads to increased viscosity and reduced polymerisation temperatures. Reducing the temperature generated during polymerisation could reduce the thermal cellular necrosis experienced in vivo, reducing the probability of aseptic loosening. Furthermore, a reduction in the exotherm of bone cement will reduce residual stresses within the cement mantle as a consequence of excessive shrinkage. Superior mechanical performance has been shown through the inclusion of MWCNTs due to their high aspect ratios and enhanced mechanical properties (*cf.* carbon fibres) [67]. Furthermore, the use of chemically functionalised MWCNT has also been shown to potentially increase nanotube dispersion and improve the interface between the host PMMA matrix and the nanotubes [68]. The use of CNT in cemented TJRs may also offer additional biological benefits such as biosensing, controlled drug release and stimulation of bone regrowth. CNTs, due to the presence of van der Waals forces, exhibit a tendency to aggregate into large bundles that could be potentially detrimental to service life of the bone cement. Enhanced fatigue performance has previously been cited for PMMA bone cement with a high degree of MWCNT dispersion; however the mixing techniques utilised were not clinically applicable, highlighting the potential value of developing a clinically transferable CNT-reinforced PMMA bone cement.

References

1. Ajayan PM (1997) Carbon nanotubes: novel architecture in nanometer space. Prog Cryst Growth Characterization Mater 34(1–4):37–51
2. Amstutz HC, Grigoris P, Dorey FJ (1998) Evolution and future of surface replacement of the hip. J Orthop Sci 3:169–186
3. Andrews R, Jacques D, Qian DL, Rantell T (2002) Multiwall carbon nanotubes: synthesis and application. Acc Chem Res 35(12):1008–1017
4. Andrews R, Weisenberger MC (2004) Carbon nanotube polymer composites. Curr Opin Solid State Mater Sci 8(1):31–37
5. Apple D (2003) A pioneer in the quest to eradicate world blindness. Bull World Health Organ 81(10):756–757
6. Baleani M, Cristofolini L, Minari C, Toni A (2003) Fatigue strength of PMMA bone cement mixed with gentamicin and barium sulphate vs pure PMMA. Proceedings of the Institution of Mechanical Engineers, Part H. J Eng Med 217:9–12
7. Baughman RH, Zakhidov AA, De Heer WA (2002) Carbon nanotubes—the route toward applications. Science 297(5582):787–792
8. Bellare A, Fitz W, Gomoll A, Turell MB, Scott R, Thornhill T (2002) Using nanotechnology to improve the performance of acrylic bone cements. Orthop J Harvard Medical School 4:93–96
9. Blakslee OL, Proctor DG, Seldin EJ, Spence GB, Weng T (1970) Elastic constants of compression-annealed pyrolytic graphite. J Appl Phys 41(8):3373–3382
10. Bonnet P, Sireude D, Garnier B, Chauvet O (2007) Thermal properties and percolation in carbon nanotube-polymer composites. Appl Phys Lett 91:201910-1-3

11. ISO, Internal Standards Organisation (2003) Implants for surgery - Acrylic resin cements. ISO 5833:2002
12. Callister WD (2000) Chapter 17: Composites materials science and engineering: an introduction. Wiley, New York, pp 520–561
13. Charnley J (1960) Anchorage of the femoral head prosthesis to the shaft of the femur. J Bone Joint Surg 42B(1):28–30
14. Chandler M, Kowalski RSZ, Watkins ND, Briscoe A, New AMR (2006) Cementing techniques in hip resurfacing. Proceedings of the Institution of Mechanical Engineers Part H. J Eng Med 220(H2):321–331
15. Chun AL, Moralez JG, Fenniri H, Webster TJ (2004) Helical rosette nanotubes: a more effective orthopaedic implant material. Nanotechnology 15(4):234–239
16. Colbert DT (2003) Single-wall nanotubes: a new option for conductive plastics and engineering polymers. Plast Addit Compounding 5(1):18–25
17. Cui D, Tian F, Ozkan CS, Wang M, Gao H (2005) Effect of single wall carbon nanotubes on human HEK293 cells. Toxicol Lett 155(1):73–85
18. Demczyk BG, Wang YM, Cumings J, Hetman M, Han W, Zettl A, Ritchie RO (2002) Direct mechanical measurement of the tensile strength and elastic modulus of multiwalled carbon nanotubes. Mater Sci Eng A 334(1–2):173–178
19. Demian HW, McDermott K (1998) Regulatory perspective on characterization and testing of orthopedic bone cements. Biomaterials 19(17):1607–1618
20. Dunne NJ, Orr JF (2002) Curing characteristics of acrylic bone cement. J Mater Sci Mater Med 13(1):17–22
21. Dunne NJ, Orr JF, Mushipe MT, Eveleigh RJ (2003) The relationship between porosity and fatigue characteristics of bone cements. Biomaterials 24(2):239–245
22. Dunne NJ, Ormsby R, McNally T, Mitchell CA, Martin D, Halley P, Nicholson T, Schiller T, Gahan L, Musumeci A, Smith SV (2010) Nanocomposite bone cements for orthopaedic applications. J Biomech 43(1):S53
23. Eriksson AR, Alberksson T (1983) Temperature threshold levels for heat-induced bone tissue injury—a vital microscopy in the rabbit. J Prosthet Dent 50(1):101–107
24. Fadeel B, Kagan V, Krug H, Shvedova A, Svartengren M, Tran L, Wiklund L (2007) There's plenty of room at the forum: potential risks and safety assessment of engineered nanomaterials. Nanotoxicology 1(2):73–84
25. Feith R (1975) Side effect of acrylic cement implanted into bone. Acta Orthop Scand 214:161
26. Franklin P, Wood DJ, Bubb NL (2005) Reinforcement of poly(methyl methacrylate) denture base with glass flake. Dent Mater 21(4):365–370
27. Fukushima H, Hashimoto Y, Yoshiya S, Kurosaka M, Matsuda M, Kawamura S, Iwatsubo T (2002) Conduction analysis of cement interface temperature in total knee arthroplasty. Kobe J Med Sci 48:63–72
28. Furnes O, Havelin LI, Espehaug B, Fenstad AM, Steindal K (2005) The Norwegian Arthroplasty Register, Annual Report, Department of Orthopaedic Surgery, Haukeland University Hospital. http://wwwhaukelandno/nrl/
29. Gilbert JL, Ney DS, Lautenschlager EP (1995) Self-reinforced composite poly(methyl methacrylate): static and fatigue properties. Biomaterials 16(14):1043–1055
30. Gilbert JL, Hasenwinkel JM, Wixson RL, Lautenschlager EP (2000) A theoretical and experimental analysis of polymerisation shrinkage of bone cement: a potential major source of porosity. J Biomed Mater Res 52(1):210–218
31. Gojny FH, Nastalczyka J, Roslaniec Z, Schulte K (2003) Surface modified multi-walled carbon nanotubes in CNT/epoxy-composites. Chem Phys Lett 370(5–6):820–824
32. Granchi D, Cenni E, Savarino L, Ciapetti G, Forbicini G, Vancini M, Maini C, Baldini N, Giunti A (2002) Bone cement extracts modulate the osteoprotegerin/osteoprotegerin-ligand expression in Mg63 osteoblast-like cells. Biomaterials 23(11):2359–2365
33. Harper EJ, German MJ, Bonfield W, Braden M, Dingeldein E, Wahlig H (2000) A comparison of VersaBondTM, a modified acrylic bone cement with improved handling properties, to Palacos R, Simplex P Trans. Sixth World Biomater Cong, Kamuela (Big Island). p540

34. Hendriks JGE, van Horn JR, van der Mei HC, Busscher HJ (2004) Backgrounds of antibiotic-loaded bone cement and prosthesis-related infection. Biomaterials 25(3):545–556
35. Herzog E, Casey A, Lyng FM, Chambers G, Byrne HJ, Davoren M (2007) A new approach to the toxicity testing of carbon-based nanomaterials—the clonogenic assay. Toxicol Lett 174:49–60
36. Hrkach JS, Peracchia MT, Bomb A, Lotan N, Langer R (1997) Nanotechnology for biomaterials engineering: structural characterization of amphiphilic polymeric nanoparticles by ¹H NMR spectroscopy. Biomaterials 18(1):27–30
37. Huang KY, Yan JJ, Lin RM (2005) Histopathologic findings of retrieved specimens of vertebroplasty with polymethylmethacrylate cement—case control study. Spine 30(19): E585–E588
38. Hull D, Clyne TW (1996) An introduction to composite materials. Cambridge University Press, Cambridge
39. Hungerford DS, Jones LC (1988) The rationale of cementless revision of cemented arthroplasty failures. Clin Orthop Relat Res 235:12–24
40. Iijima S (1991) Helical microtubules of graphitic carbon. Nature 354:56–58
41. Iijima S (2002) Carbon nanotubes: past, present, and future. Physica B Condens Matter 323(1–4):1–5
42. Kagan VE, Tyurina YY, Tyurin VA, Konduru NV, Potapovich AI, Osipov AN (2006) Direct and indirect effects of single walled carbon nanotubes on RAW 2647 macrophages: role of iron. Toxicol Lett 165(1):88–100
43. Karrholm J, Garellick G, Rogmark C, Herberts P (2008) Annual Report 2007 The Swedish National Hip Arthroplasty Register. http://wwwjruorthopguse
44. Keener JD, Callaghan JJ, Goetz DD, Pederson DR, Sullivan PM, Johnston RC (2003) Twenty-five-year results after charnley total hip arthroplasty in patients less than fifty years old—a concise follow-up of a previous report. J Bone Joint Surg Am 85A(6):1066–1072
45. Kotha SP, Li C, Schmid SR, Mason JJ (2004) Fracture toughness of steelfiber—reinforced bone cement. J Biomed Mater Res A 70(3):514–521
46. Kotha SP, Li C, McGinn P, Schmid SR, Mason JJ (2006) Improved mechanical properties of acrylic bone cement with short titanium fiber reinforcement. J Mater Sci Mater Med 17(8):743–748
47. Kuehn KD, Ege W, Gopp U (2005) Acrylic bone cements: compositions and properties. Orthop Clin North Am 36:17–28
48. Lee RR, Ogiso M, Watanabe A, Ishihara K (1997) Examination of hydroxyapatite filled 4-META/MAA-TBB adhesive one cement in vitro and in vivo environment. J Biomed Mater Res 38(1):11–16
49. Lennon AB, Prendergast PJ (2002) Residual stress due to curing can initiate damage in porous bone cement: experimental and theoretical evidence. J Biomech 35(3):311–321
50. Lewis G (1997) Properties of acrylic bone cement: state of the art review. J Biomed Mater Res 38(2):155–182
51. Lewis G (2000) Relative roles of cement molecular weight and mixing method on the fatigue performance of acrylic bone cement: simplex-p versus osteopal. J Biomed Mater Res 53(1):119–130
52. Lewis G, Janna S, Carroll M (2003) Effect of test frequency on the in vitro fatigue life of acrylic bone cement. Biomaterials 24(6):1111–1117
53. Linder L (1977) Reaction of bone to the acute chemical trauma of bone cement. J Bone Joint Surg 59(1):82–87
54. Mandell JF, Huang DD, McGarry FJ (1980) Crack propagation modes in injection molded fibre reinforced thermoplastics. School of Engineering, Massachusetts Institute of technology, Cambridge, Massachusetts, pp 1–34
55. Marrs B, Andrews R, Rantell T, Pienkowski DA (2005) Augmentation of acrylic bone cement with multiwall carbon nanotubes. Part A. J Biomed Mater Res 77(A):269–276
56. Marrs B (2007) Carbon nanotube augmentation of a bone cement polymer. PhD Thesis, University of Kentucky

57. Martini FH, Bartholomew EF (2000) Chapter 6: the skeletal system. Essentials of anatomy & physiology, 2nd edn. Prentice-Hall, Inc, New Jersey, pp 120–165
58. Mjoberg B, Petterson H, Roseqvist R, Rydholm A (1984) Bone cement, thermal injury and radiolucent zone. Acta Orthop Scand 55:597–600
59. Moritz AR, Henriques FC (1947) Studies of thermal injury—the relative importance of time and surface temperature in the causation of cutaneous burns. Am J Pathol 23:695–720
60. Morscher EW, Wirz D (2002) Current state of cement fixation in THR. Acta Orthop Belg 68(1):1–12
61. Mousa WF, Kobayashi M, Shinzato S, Kamimura M, Neo M, Yoshihara S, Nakamura T (2000) Biological and mechanical properties of PMMA-based bioactive bone cements. Biomaterials 21(21):2137–2146
62. Murphy BP, Prendergast PJ (2000) On the magnitude and variability of the fatigue strength of acrylic bone cement. Int J Fatigue 22(10):855–864
63. Narva KK, Lassila LVJ, Vallittu PK (2005) Flexural fatigue of denture base polymer with fiber-reinforced composite reinforcement. Compos Part A Appl Sci Manuf 36(9):1275–1281
64. Norman TL, Kish V, Blaha JD, Gruen TA, Hustosky K (1995) Creep characteristics of hand-mixed and vacuum-mixed acrylic bone-cement at elevated stress levels. J Biomed Mater Res 29(4):495–501
65. NJR (2006) National joint registry for England and Wales 3rd annual clinical report January 2007, from wwwnjrcentreorguk
66. Nuno N, Amabili M (2002) Modelling debonded stem-cement interface for hip implants: effects of residual stresses. Clin Biomech 17(1):41–48
67. Ormsby R, McNally T, Mitchell CA, Dunne N (2010) Incorporation of multiwall carbon nanotubes to acrylic based bone cements: effects on mechanical and thermal properties. J Mech Behav Biomed Mater 3(2):136–145
68. Ormsby R, McNally T, Mitchell CA, Dunne N (2010) Influence of multiwall carbon nanotube functionality and loading on mechanical properties of PMMA/MWCNT bone cements. J Mater Sci Mater Med 21:2287–2292
69. Orr JF, Dunne NJ, Quinn JC (2003) Shrinkage stresses in bone cement. Biomaterials 24(17): 2933–2940
70. Pal S, Saha S (1982) Stress relaxation and creep behaviour of normal and carbon fibre reinforced acrylic bone cement. Biomaterials 3:93–96
71. Pande S, Mathur RB, Singh BP, Dhami TL (2009) Synthesis and characterization of multi-walled carbon nanotubes-polymethyl methacrylate composites prepared by in situ polymerization method. J Polym Compos 30(9):1312–1317
72. Pilliar RM, Blackwell R, Macnab I, Cameron HU (1976) Carbon fiber reinforced bone cement in orthopedic surgery. J Biomed Mater Res 10(6):893–906
73. Prendergast PJ (2001) Chapter 35: Bone prostheses and implants. Bone mechanics handbook Cowin, SC. CRC Press, LLC
74. Pukanszky B (2005) Interfaces and interphases in multicomponent materials: past, present, future. Eur Polym J 41(4):645–662
75. Robinson RP, Wright TM, Burstein AH (1981) Mechanical properties of poly(methyl methacrylate) bone cements. J Biomed Mater Res 15(2):203–208
76. Roques A, Browne M, Taylor A, New A, Baker D (2004) Quantitative measurement of the stresses induced during polymerisation of bone cement. Biomaterials 25(18):4415–4424
77. Saha S, Pal S (1986) Mechanical characterization of commercially made carbonfibre-reinforced polymethylmethacrylate. J Biomed Mater Res 20(6):817–826
78. Sauer JA, Richardson GC (1980) Fatigue of polymers. Int J Fract 16(6):499–532
79. Scheirs J (2000) Chapter 15: Failure of fibre-reinforced composites. Compositional and failure analysis of polymers a practical approach. Wiley, New York, pp 449–481
80. Scheirs J (2000) Chapter 12: Mechanical failure mechanisms of polymers. Compositional and failure analysis of polymers a practical approach. Wiley, New York, pp 304–362
81. Serbetci K, Korkusuz F, Hasirci N (2004) Thermal and mechanical properties of hydroxyapatite impregnated acrylic bone cements. Polym Test 23(2):145–155

82. Shannon BD, Klassen JF, Rand JA, Berry DJ, Trousdale RT (2003) Revision total knee arthroplasty with cemented components and uncemented intramedullary stems. J Arthroplasty 18(7):27–32

83. Shvedova AA, Kisin ER, Mercer R, Johnson VJ, Potapovich AI (2005) Unusual inflammatory and fibrogenic pulmonary responses to single-walled carbon nanotubes in mice. Am J Physiol 289:L698–L708

84. Shvedova AA, Fabisiak JP, Kisin ER, Murray AR, Roberts JR, Tyurina YY (2008) Sequential exposure to carbon nanotubes and bacteria enhances pulmonary inflammation and infectivity. Am J Respir Cell Mol Biol 38(5):579–590AA

85. Shvedova AA, Kisin ER, Porter D, Schulte P, Kagan VE, Fadeel B, Castranova V (2009) Mechanisms of pulmonary toxicity and medical applications of carbon nanotubes: two faces of Janus? Pharmacol Ther 121(2):192–204

86. Shinzato S, Kobayashi M, Mousa F, Kamimura M, Neo M, Kitamura Y, Kokubo T, Nakamura T (2000) Bioactive polymethyl methacrylate-based bone cement: comparison of glass beads, apatite- and wollastonite-containing glass ceramic, and hydroxyapatite fillers on mechanical and biological properties. J Biomed Mater Res 51(2):258–272

87. Sinha N, Yeow JTW (2005) Carbon nanotubes for biomedical applications. IEEE Trans Nanobioscience 4(2):180–195

88. Smart SK, Cassady AI, Lu GQ, Martin DJ (2006) The biocompatibility of carbon nanotubes. Carbon 44(6):1034–1047

89. Stanczyk M, van Rietbergen B (2004) Thermal analysis of bone cement polymerisation at the cement-bone interface. J Biomech 37(12):1803–1810

90. Starke GR, Birnie C, van den Blink PA (2001) Numerical modelling of cement polymerisation and thermal bone necrosis. Computer methods in biomechanics and biomedical engineering. Gordon & Breach, London

91. Stolk J, Verdonschot N, Murphy BP, Prendergast PJ, Huiskes R (2004) Finite element simulation of anisotropic damage accumulation and creep in acrylic bone cement engineering. Fract Mech 71(4–6):513–528

92. The Royal Society and the Royal Academy of Engineering (2004) Nanoscience and nanotechnologies: opportunities and uncertainties. The Royal Society and the Royal Academy of Engineering, London

93. Toksvig-Larse S, Franze H, Ryd L (1991) Cement interface temperature in hip arthroplasty. Acta Orthop 62(2):102–105

94. Topoleski LDT, Ducheyne P, Cuckler JM (1993) Microstructural pathway of fracture in poly(methyl methacrylate) bone-cement. Biomaterials 14(15):1165–1172

95. Verdonschot N, Huiskes R (1997) Acrylic cement creeps but does not allow much subsidence of femoral stems. J Bone Joint Surg Br 79B(4):665–669

96. Webster T, Waid MC, McKenzie JL, Price RL, Ejiofor JU (2004) Nano-biotechnology: carbon nanofibres as improved neural and orthopaedic implants. Nanotechnology 15(1):48–54

97. Wong EW, Sheehan PE, Lieber CM (1997) Nanobeam mechanics: elasticity, strength and toughness of nanorods and nanotubes. Science 277(5334):1971–1975

98. Wright TM, Robinson RP (1982) Fatigue crack propagation in polymethylmethacrylate bone cements. J Mater Sci 17(9):2463–2468

99. Yang J-M, Huang P-Y, Yang M-C, Lo SK (1997) Effect of MMA-g- UHMWPE grafted fiber on mechanical properties of acrylic bone cement. J Biomed Mater Res 38(4):361–369

100. Zhao B, Hu H, Mandal SK, Haddon RC (2005) A bone mimic based on the self-assembly of hydroxyapatite on chemically functionalized single-walled carbon nanotubes. Chem Mater 17(12):3235–3241

Chapter 9
Exploring the Future of Hydrogels in Rapid Prototyping: A Review on Current Trends and Limitations

Thomas Billiet, Mieke Vandenhaute, Jorg Schelfhout, Sandra Van Vlierberghe, and Peter Dubruel

9.1 Introduction

To date, organ and tissue transplantation remains one of the most important straightforward options in order to restore or enhance life expectancy. The most recent annual report prepared by the Scientific Registry of Transplant Recipients (SRTR) in collaboration with the Organ Procurement and Transplantation Network (OPTN) registered 100,597 patients in the USA awaiting transplantation at the end of 2008, while only 27,281 transplantations were performed [1]. If we keep the steady increase in life expectancy in mind, these numbers emphasize the shortage of organ donors. In addition, diseases, infections and rejection of the tissue by the host often complicate transplantation [2]. To overcome these problems associated with transplantation, in the last few decades, tissue engineering (TE) has grown as a new inter- and multidisciplinary scientific field [3]. This discipline has rapidly emerged and combines the principles of engineering and life sciences. It holds as main objective the recovery, maintenance and improvement of tissue performance [3–5]. The European Commission on Health and Consumer Protection defines TE as the persuasion of the body to heal itself through the delivery, to the appropriate site, independently or in synergy, of cells, biomolecules and supporting structures [6].

Researchers will strive to fulfil those aforementioned objectives through the utilization of isolated cells [7–10], tissue-inducing substances [11–13] and/or scaffolds [2, 3, 5, 14]. However, conventionally, the application of a supporting scaffold is preferred in circumstances where the defect acquires certain dimensions. Post-processing cell seeding and maturation to tissue has therefore been implemented as

T. Billiet • M. Vandenhaute • J. Schelfhout • S.V. Vlierberghe • P. Dubruel (✉)
Polymer Chemistry & Biomaterials Research Group, Ghent University,
Krijgslaan 281 S4 Bis, Ghent 9000, Belgium
e-mail: thomas.billiet@ugent.be; mieke.vanderhaute@ugent.be; jorg.schelfhout@ugent.be; sandra.vanvlierberghe@ugent.be; peter.dubruel@ugent.be

I. Antoniac (ed.), *Biologically Responsive Biomaterials for Tissue Engineering*,
Springer Series in Biomaterials Science and Engineering 1,
DOI 10.1007/978-1-4614-4328-5_9, © Springer Science+Business Media New York 2013

a commonly applied TE strategy [14–18]. Expanding the cell population and maturation to tissue is performed in bioreactors, which can be described as devices in which biological and/or biochemical processes are manipulated through close control of environmental and process-bound factors such as pH, temperature, pressure and nutrient and waste flow [19]. When working with low-water-content polymers, post-processing cell seeding is the only available seeding mechanism. However, insufficient cell seeding and a non-uniform cell distribution have been reported when using this methodology [19, 20]. Thus, there is a need for better and more uniform seeding mechanisms. Enhancing the seeding efficiency can be accomplished by cell encapsulation strategies, demanding a high-water-content environment.

Hydrogels based on natural or synthetic polymers have been of great interest regarding cell encapsulation [21–36]. For the past decade, such hydrogels have become especially attractive as matrices for regenerating and repairing a wide variety of tissues and organs [6, 11, 32–91]. Depending on the hydrophilicity, they can absorb up to thousands of times of their dry weight and form chemically stable or (bio)degradable gels. Depending on the nature of the hydrogel network, "physical" and "chemical" gels can be distinguished. Hydrogels are called "physical" when the network formation is reversible, in contrast to "chemical" hydrogels which are established by irreversible, covalent crosslinks. Combinations of both physical and chemical networks can also be achieved, e.g. gelatine modified with methacrylamide groups [92].

The characteristic properties of hydrogels make them especially appealing for repairing and regenerating soft tissue [31, 36–38, 84–91, 93–96]. One of the main disadvantages of processing hydrogels is the difficulty to shape them in predesigned geometries. This article provides a detailed overview of the different rapid prototyping techniques that are compatible with hydrogel manufacturing and allow to accurately shape external and internal geometries. Since we did not find an article that summarizes the potential advantages and disadvantages regarding the processing of hydrogels with RP techniques, it is the purpose to highlight the advantages, but more importantly also the current limitations, of the distinctive techniques, together with the respective hydrogels used so far.

In the first part, an introduction in scaffolding and basic concepts of scaffold-based and scaffold-free TE is given. The next part handles hydrogel-friendly RP techniques used in scaffold-based TE. Finally, the implementation of RP technology in scaffold-free TE is explained.

9.2   ECM Mimetics: Current Concepts

9.2.1   Scaffold-Based vs. Scaffold-Free TE

From a cell biology perspective, 2D cell culture models only provide physiologically compromised cells induced by an unnatural environment [97], and the lack of a 3D structure will cause cells to form a random 2D monolayer [16, 18]. In vivo, cells are subjected to growth in three dimensions and complex cell–cell interactions. This observation encouraged a paradigm shift from conventional 2D cell culture models towards 3D micro-environments [98]. To obtain a more realistic understanding of cell–cell and cell–biomaterial interactions, Kirkpatrick et al. [99] proposed the use of co-culture models in vitro. Independent of the applied strategy, the ultimate goal of TE remains the same. Nevertheless, regarding the aspect of 3-dimensional cell migration, proliferation and differentiation behaviour and requisites, one can distinguish two major premises. Currently both of them are being heavily explored. The first one is based on the presumption that cells require an assisting 3D biomaterial scaffold that closely mimics the corresponding extracellular matrix (ECM) [98, 100]. In this approach, the biomaterial construct acts as a necessary cell guide and supporting template. The second one finds its roots in the hypothesis that cells have a considerable potency to self-organize through cell–cell interactions and is referred to as "scaffold-free TE" [101]. While the former theory maximizes the role of a supporting structure as a cell guide and minimizes the potency to self-assembly, the latter reverses the importance of both contributions.

9.2.2 Scaffolds

Ideally, scaffolds can be seen as ECM biomimetic structures with three main objectives [16, 17]: (1) defining a space that moulds the regenerating tissue, (2) temporary substitution of tissue functions and (3) guide for tissue ingrowth. It is clear that scaffold design should meet the needs of some basic requirements to be able to meet those objectives, including [2, 14, 16–18] the following: high porosity (preferably 100 % interconnectivity for optimal nutrient/waste flow and tissue ingrowth); relevant geometry and pore dimensions (five to ten times the cell diameter); biodegradable with adjusted degradation time; maintaining the mechanical integrity during a prefixed time frame; it should have suitable cell–biomaterial interactions and it should be easy to manufacture. Adjusting the mechanical and degradation properties to the desired tissue is essential. Either enzymatic or non-enzymatic hydrolytic processes control the degradation profile. Specifically, TE requires biomaterials that provoke cell interactions (~bioactivity) [102] and as little as possible adverse body reactions (~biocompatibility) [103]. Control over the material bioactivity can be achieved by incorporating growth factors [104], enzymatic recognition sites [105], adhesion factors [93, 106] or material modifications [105]. Material modification is a general term indicating either bulk modification [102, 107] or surface modification [102, 108, 109]. Modifying the bulk properties is closely related to material biocompatibility, the physical and chemical properties covering the lifespan of the implant [110], while varying the surface chemistry reflects on the initial cell/tissue–material interactions [110, 111].

Figure 9.1 illustrates schematically the complex multidisciplinary interactions inherent towards scaffold fabrication. In the subscience of scaffolding,

Fig. 9.1 Schematic illustration integrating the complex multi-disciplinary needs which determine the constraints for the ideal scaffold fabrication design

both conventional and rapid prototyping (RP) techniques have been explored. Conventional scaffold fabrication set-ups include techniques such as particulate leaching [84, 112–114], gas foaming [113–116], fibre networking [117, 118], phase separation [119, 120], melt moulding [121, 122], emulsion freeze drying [123, 124], solution casting [125, 126], freeze drying [80, 86, 127] and combinations of those. Conventional/classical approaches are defined as processes that create scaffolds with a continuous, uninterrupted pore network. Nonetheless, they completely lack long-range micro-architectural channels [18]. Other reported disadvantages involve low and inhomogeneous mechanical strength, limited porosity and insufficient interconnectivity, inability to spatially design the pore distribution (internal channels) and pore dimensions and difficulty in manufacturing patient-specific implants (control over external geometry is limited) [18, 128]. Furthermore, the use of organic solvents during processing is seen as a second major drawback in addition to the above-mentioned architectural drawbacks. The presence of organic solvent residues can pose significant constraints related to toxicity risks and carcinogenetic effects [18]. Despite some adaptations over the years the scaffold design remains process dependent by means of classical approaches. A more design-dependent method would be attractive, and this can be attained by RP techniques.

9.2.3 Controlling the External and Internal Geometry

Solid freeform fabrication (SFF) is the general term covering all techniques that produce objects through sequential delivery of energy and/or material. When rapid fabrication of a prototype, a finished object or a tool is pursued, they are, respectively, called rapid prototyping, rapid manufacturing (RM) and rapid tooling (RT) [129]. By means of RP, an additive computer-controlled layer-by-layer process generates a scaffold. 3D computer models shape the external design, and such models can be designed either by CAD software or by modelling imaging data (CT, MRI). On the other hand, the internal architecture is determined by the processing of the CAD data into an STT file and subsequent slicing of the STT data (generation of the machine parameters). This directly indicates one of the greatest assets of RP: direct fabrication of scaffolds with a complex, patient-specific external geometry [16, 18, 64] in combination with a precise control over the internal architecture (limited by the resolution of the system). Other advantages comprise high degree of interconnectivity, possibility to use heterogeneous materials, high speed due to a high degree of automization and the limited number of process steps, and a superior cost-efficiency [2, 64]. Both direct and indirect RP methods exist. In the former case, the scaffold is directly processed from a biomaterial, and in the latter case the scaffold is processed out of an RP mould. Worldwide more than 30 different RP techniques are being applied in the most diverse industries [129], and around 20 of them found applications in the biomedical field. This particular subject has been reviewed by several authors [2, 129, 130]. Despite the wide diversity of RP technologies, only some of them seem to be compatible for the processing of hydrogels. The next chapter describes the different RP scaffolding techniques compatible with hydrogels.

9.3 Rapid Prototyping Hydrogels: Powerful Aid in Making Scaffold-Based Tissue Engineering Work

A primary classification of the SFF techniques supporting biomedical applications can be made hinged on the working principle: (1) laser-based, (2) nozzle-based and (3) printer-based systems. Laser-based systems benefit from the photopolymerization pathway as a basis to fabricate cross-linked polymeric TE scaffolds. The well-known processing of (pre)polymers by dint of extrusion/dispension supports the second category of RP systems. The last subclass works with powder beds and deposition of a binder that fuses the particles, or direct deposition of material using inkjet technology. An important characteristic feature of every technique will be its

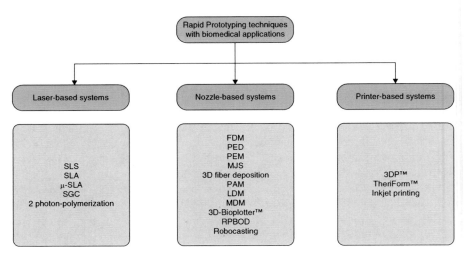

Fig. 9.2 Classification of RP techniques with biomedical applications into laser-, nozzle-, and printer-based systems

resolution. Every technique is subdued to a lower technical limit size of the smallest details producible. This so-called lower limit shows a clear relationship with the feasible scale of the object: the higher the resolution of the smallest details, the smaller will be its maximum object size [131]. However, since not all RP techniques are applicable for the processing of hydrogel materials, some more than others, the amount of RP technologies is further diluted. Figure 9.2 classifies the different RP techniques with biomedical applications. The fabrication of hydrogel scaffolds requires mild processing conditions. Some of the techniques mentioned in Fig. 9.2 are not able to meet those constraints due to rather harsh processing conditions. Explaining those is not the purpose of this review. Only the hydrogel-compatible systems are explained in detail.

9.3.1 Laser-Based Systems

9.3.1.1 Working Principles and Recent Trends of Laser-Based Systems

With the exception of selective laser sintering (SLS), all of the laser-based systems are suitable for hydrogel processing. Unlike the nozzle- and printer-based systems that sequentially deposit material, this subclass sequentially deposits light energy in specific predefined patterns. This directly implies that only photocrosslinkable pre-polymers can be employed to finally obtain a crosslinked hydrogel network.

Stereolithography (SLA). SLA is considered to be the first commercially available SFF technique, developed by 3D systems in 1986 [131]. An SLA apparatus (Fig. 9.3) consists of a reservoir to be filled with a liquid photocurable resin, a laser source

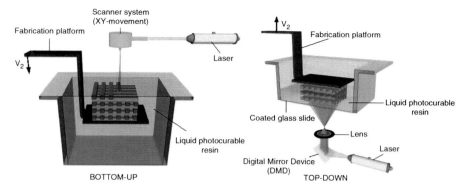

Fig. 9.3 Scheme of bottom-up (*left*) and top-down (*right*) stereolithography set-ups. The bottom-up set-up shown, is an example of a system whereby the laser scans the surface for the curing of the photosensitive material. In the example of the top-down set-up, dynamic light projection technology is used to cure a complete 2D layer at once

(commonly UV light), a system that controls the xy-movement of the light beam and a fabrication platform that permits movement in the vertical plane. Scanning the surface of the photosensitive material produces 2D patterns of polymerized material through single photon absorption at the surface of the liquid material. The build-up of a 3D construct is made possible using a layer-by-layer approach, whereby the fabrication platform moves stepwise in the z-direction after a 2D layer is finished. The step height of the fabrication platform is typically smaller than the curing depth, ensuring good adherence of subsequent layers. Post-treatment steps involving washing off excess resin and further curing with UV light are in most cases necessary. Arcaute et al. [132] demonstrated the possibility to alter the resolution of the cure depth by varying the laser energy, the concentration of poly(ethylene glycol) dimethylacrylate (PEG-DMA) as photocrosslinkable material and the type and concentration of the photoinitiator. In addition, adjusting the scanning speed influences the cure depth.

Micro-stereolithography (μ-SLA). The working mechanism of μ-SLA can be considered the same as that of a normal SLA. The difference between both involves the resolution of the system. μ-SLA systems are typically able to build very accurately (a few microns) objects of several cubic centimetres [133–136].

Figure 9.3 (left) shows a scheme of the so-called bottom-up set-up, in which an object is built from a fabrication support just below the resin surface. Subsequent layers are being cured on top of the previous layers by irradiation from above. Although to date this is the most applied set-up [132, 134–139], a top-down approach (Fig. 9.3 right) is gaining interest [70, 131, 133]. Top-down set-ups have a non-adhering, transparent plate acting as the bottom of the liquid reservoir. Polymerization of the photosensitive material occurs through irradiation from underneath, and the fabrication platform moves in the opposite direction as in the bottom-up approach. In this way, every newly formed layer is located beneath the previous one. Separating every newly formed layer from the bottom plate will subject the structure to larger mechanical forces but on the other hand the vat content can be minimized, the irradiated surface will not be exposed to the atmosphere (crosslinking efficiency limitation due to oxygen will be minimized), recoating the structure with a new resin layer is not required and the illuminated area is always smooth [131]. Another recent and more fundamental trend in the field of SLA is the emerging use of digital light projection (DLP) technology [133, 139–141]. The working principle is illustrated in Fig. 9.3 (right) in the top-down scheme, but is also applicable for bottom-up set-ups. Projection technology enables the curing of a complete layer of resin in one go, which obviously reflects on the building time. A Digital Micro-mirror Device™ (DMD) consists of an array of mirrors, which can independently be tilted in an on/off state. In this way the DMD serves as a dynamic mask that projects a 2D pattern (often designed in PowerPoint slides) on the surface. Instead of DMD, LCD displays have also been employed as a dynamic mask projector [142, 143]; however, DMD offers better performance in terms of optical fill factor and light transmission [139].

Solid ground curing (SGC). A projection technology somewhat similar to DMD technology is SGC developed by Cubital Inc. [144, 145]. Coating the fabrication platform by spraying a photosensitive resin is the first step in the SGC workflow. Meanwhile, the machine prints a photomask of the layer to be built on a glass plate above the fabrication platform. The printing process resembles the one applied in commercial laser printers. Solidification of the sprayed layer occurs when the mask is exposed to UV light, only permitting irradiation of the transparent regions. After the layer is completed, excess liquid resin is removed by vacuum and replaced by a wax to support the next layer. Before repeating the cycle, the layer is milled flat to an accurate, reliable finish for the next layer.

Two-photon polymerization (2PP). 2PP is an emerging state-of-the-art laser-based technique. In this process, light is used to trigger a chemical reaction leading to polymerization of a photosensitive material. Unlike other light curing systems (single photon polymerization), 2PP initiates the polymerization through radiation with near-infrared femtosecond laser pulses of 800 nm (Fig. 9.4). In the focal point, a suitable photoinitiator absorbs two photons, with a wavelength of 800 nm,

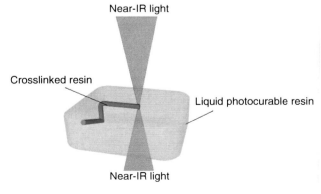

Fig. 9.4 Working principle of two-photon photopolymerization. In the focal point of the near-IR laser beams, the photosensitive polymer is crosslinked. A "true" 3D object is obtained

simultaneously, causing them to act as one photon of 400 nm, and thus starting the polymerization reaction [133]. The non-linear excitation nature triggers polymerization only in the focal point, while other regions remain unaffected. This approach has potential solidification resolutions below the diffraction limit of the applied light.

Moving the laser focus enables the fabrication of a direct "true" 3D object into the volume of the photosensitive material. Creating reproducible micron-sized objects with feature sizes of less than 100 nm [146] is attainable, thus being superior to all other SFF techniques regarding accuracy and resolution.

9.3.1.2 Current Limitations and Hydrogel Feasibility for Laser-Based Systems

SLA has already been frequently applied to develop porous hydrogel-based scaffolds (Table 9.1). Yu et al. [147] have described the patterning of 2-hydroxyethyl methacrylate (HEMA) followed by drying and subsequent rehydration to enable cell adhesion. Initially, the procedure was applied to create single-layer structures; however, at present, multiple layers can be superimposed to generate porous 3D scaffolds. Liu et al. [30] selected poly(ethylene glycol)diacrylate (PEGDA) hydrogels to construct 3D scaffolds layer by layer using emulsion masks. In that study, three hydrogel layers were fused at a resolution of several hundreds of microns. Interestingly, the pores were interconnected enabling cell survival through convective flow of culture medium. However, this procedure is time consuming, requires a great number of prefabricated masks depending on the required shape and is not completely automated. Therefore, one can discuss whether this can be considered as a genuine rapid prototyping technique. Lu et al. [139, 148] have adopted the above-mentioned procedure to develop scaffolds possessing complex internal architectures

Table 9.1 Hydrogel materials explored in laser-based systems

Laser-based systems	Hydrogel materials	Cell encapsulation	References
SLA	Gelatin-methacrylate/ gelatin-methacrylamide	×	[70, 133, 159]
	Hyaluronic acid-methacrylate	√	[154, 156]
	Cysteine-modified agarose	×	[155]
	HEMA	×	[147]
	PEG-D(M)A	√	[30, 31, 132, 137, 139, 153]
	PEG-D(M)A	×	[141, 149–152]
μ-SLA	Alginate + acrylated TMC/TMP	√	[135, 136, 157]
	Gelatin-methacrylamide	×	[133]
	PEG-DA	×	[158, 164]
2PP	Gelatin-methacrylamide	×	[133]
	PEG-DA	×	[160, 161]

and spatial patterns within a Z-range of several hundreds of microns. Another research group even produced PEG-based scaffolds in the millimetre scale [132, 149–151]. Several other research groups also selected PEGDA as starting material [31, 152]. Yasar et al. [31] successfully plotted 100 μm-sized complex scaffold architectures. The swelling effects of the PEGDA, however, prevented the fabrication of highly reproducible samples below 100 μm. Higher resolutions could be obtained in the presence of UV absorbers since they prevent the internal reflection of the UV light within the polymer solution. More recently, the feasibility of plotting cell-encapsulated hydrogels has been evaluated. Several research groups have already reported on the cell encapsulation in photocrosslinkable poly(ethylene glycol) (PEG) microgels [30, 153].

In addition to synthetic polymers, photocrosslinkable biopolymers including hyaluronic acid (HA) [154] and gelatin [70] derivatives have already been printed with or without cells using SLA.

In order to enhance the cell-interactive properties of a material, different surface functionalization strategies can be elaborated [31]. Luo et al. [155] grafted RGD-containing peptide sequences on the surface of scaffolds composed of cysteine-modified agarose. Han et al. [141] applied a fibronectin coating on the surface of a PEGDA scaffold to improve the attachment of murine marrow-derived progenitor cells. However, important limitations of laser-based systems include both the need for photocrosslinkable materials as well as the effect of the applied UV light on the encapsulated cells [156].

Since shrinkage occurs after post-processing of scaffolds developed using SLA, a major drawback is its limited resolution [138]. In addition, due to scattering phenomena of the applied laser beam, a significant deformation occurs when relatively small objects are developed. The produced hydrogel is often weak upon removal and post-curing is often essential.

Therefore, μ-SLA was introduced to counter the limitations of SLA from a resolution point of view. For example, Lee et al. [135, 136, 157] developed a hybrid scaffold consisting of an acrylated trimethylene carbonate/trimethylolpropane (TMC/TMP) framework and an alginate hydrogel for chondrocyte encapsulation. The encapsulated cells retained their phenotypic expression within the structure and the scaffold remained mechanically stable up to 4 weeks after implantation in mice. Barry et al. [158] have combined direct ink writing (DIW) with in situ photopolymerization to create hydrogel scaffolds possessing micrometre-sized features. Using this approach, another research group even realized submicron range structures based on PEGDA [159].

In addition to SLA techniques, SGC also shows potential to be applied in the development of porous hydrogel-based scaffolds for tissue engineering, as already indicated before [138]. However, up to now, no literature data regarding this application and hydrogel processing can be found.

In order to achieve 3D subcellular resolution during scaffold development, 2PP can offer a suitable alternative for SLA. Since this technique was only properly introduced recently, only few reports can be found in literature regarding the application of 2PP to produce porous hydrogels.

For example, Schade et al. [160] developed hydrogel-like scaffolds possessing well-defined 3D structures using a methacrylated polyurethane and PEGDA as starting materials. Ovsianikov et al. [161] also selected PEGDA as starting material for 2PP scaffolds. More recently, they evaluated the feasibility to produce porous scaffolds using methacrylamide-modified gelatin developed in our research group [162]. The results of their study will be published in due course [163].

Table 9.5 summarizes more technical details on the implementation of hydrogels in the different RP technologies discussed.

9.3.2 Nozzle-Based Systems

9.3.2.1 Working Principles and Recent Trends of Nozzle-Based Systems

The class of nozzle-based systems is characterized by a wide diversification (Fig. 9.2). Fused deposition modelling (FDM), 3D fibre deposition, precision extrusion deposition (PED), precise extrusion manufacturing (PEM) and multiphase jet solidification (MJS) are techniques based on a melting process. Generally, the melt process involves elevated temperatures which are undesirable from the perspective of scaffold bioactivity [17]. Researchers have therefore tried to bring forth several other techniques that overcome this limitation by applying a dissolution process, which is attractive for the processing of hydrogels. Four major nozzle designs have been described in literature: pressure-actuated, solenoid-actuated, piezoelectric and volume-actuated nozzles [164, 165]. These nozzle types can be found in the following systems.

Pressure-assisted microsyringe (PAM). A technique that resembles FDM without the need for heat is the PAM technique, developed by Vozzi et al. [166]. The set-up consists of a 5–20 μm pneumatic-driven glass capillary syringe that can move in the vertical plane and deposits material on a substrate. The substrate proceeds in the planar field relative to the syringe. Transforming jpeg or bitmap images into a sequential list of linear coordinates easily allows to deposit practically any type of structure in subsequent layers [167]. Material viscosities, deposition speed, tip diameter and applied pressure correlate with the final deposited strand dimensions. The PAM system has been described in several publications [168, 169]. Recently, the fabrication of hydrogel scaffolds was successful with the PAM method [170, 171].

Low-temperature deposition modelling (LDM). Proposed by Xiong et al. [172] in 2002, LDM found its way as an RP system with biomedical applications. The key feature of this technique is a non-heating liquefying processing of materials [172]. Using temperatures below 0 °C, the material solution is solidified when deposited on the fabrication platform [173]. The material gets extruded out of a nozzle capable of moving in the *xy*-plane onto a built platform movable in the *z*-direction. Incorporating multiple nozzles with different designs into the LDM technique gave existence to multi-nozzle (low-temperature) deposition modelling (MDM, M-LDM) [165, 165, 174, 175]. A multinozzle low-temperature deposition and manufacture

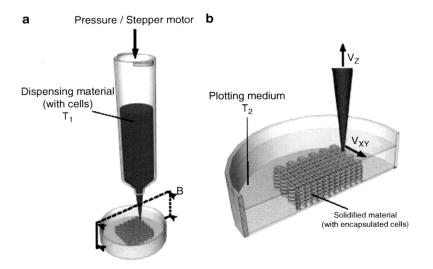

Fig. 9.5 Scheme of 3D-Bioplotter™ dispensing principle. In the 3D-Bioplotter™ system, the nozzle works pneumatically or via volume-driven injection. This also illustrates the principle of nozzle-based systems in general, where a nozzle is used for the deposition of material. Key difference with other nozzle-based systems is the ability to plot into a liquid medium with matching density, thus introducing buoyancy compensation

(M-LDM) system is proposed to fabricate scaffolds with heterogeneous materials and gradient hierarchical porous structures by the incorporation of more jetting multi-nozzle into the system [176]. Biomolecules can be applied in the LDM process to fabricate a bioactive scaffold directly.

3D-Bioplotter™. This 3D dispensing process, displayed in Fig. 9.5, has been introduced by Landers and Mülhaupt in 2000 at the Freiburg Research Centre [177]. The technique was specifically developed to produce scaffolds for soft-tissue engineering purposes, and simplifying hydrogel manufacturing. The three-dimensional construction of objects occurs in a laminar fashion by the computer-controlled deposition of material on a surface. The dispension head moves in three dimensions, while the fabrication platform is stationary. It is possible to perform either a continuous dispension of microstrands or a discontinuous dispension of microdots. Liquid flow is generated by applying filtered air pressure (pneumatic nozzle), or using a stepper-motor (volume-driven injection (VDI) nozzle). The ability to plot a viscous material into a liquid (aqueous) medium with a matching density is the key feature of this process. Low viscous materials in particular benefit from this buoyancy compensation principle. Since heating is not required, the system can process thermally sensitive biocomponents, and even cells. Curing reactions can be performed by plotting in a co-reactive medium or by two-component dispensing using mixing nozzles. The strand thickness can be modulated by the varying material viscosity, deposition speed (speed in the planar field), tip diameter or applied pressure. Constructs built by this plotting technique mostly have smooth strand surfaces, which are not desired for appropriate cell attachment. Therefore, further surface treatment is required to render the surface favourable for cell adhesion. Recently Kim et al. [178] adapted the Bioplotting device with a piezoelectric transducer (PZT) generating vibrations while plotting PCL. Scaffolds built had a rougher surface and showed better cell adhesion than the ones built with the conventional set-up.

Rapid prototyping robotic dispensing (RPBOD). Ang et al. [179] adopted an analogous concept of the 3D-Bioplotter™ technology to develop a robotic dispensing system: the RPBOD system. The set-up consists of a one-component pneumatic dispenser.

Robocasting. The laminar deposition of highly loaded ceramic slurries (typically 50–65 vol.% ceramic powder) to build a 3D construct using robotics is called robocasting [180]. Unlike the bioplotting process, in most cases the robocasting set-up has a stationary dispension head, while the fabrication platform moves in the planar and vertical field. Inks used for robocasting have to flow under stress and recover enough stiffness such that, when the stress is released, they can bear both the extruded filament weight and the weight of successive layers [181, 182]. Thus, in contrary to other SFF techniques, the concept of robocasting relies essentially on the rheology of the slurry and also on the partial drying of the deposited layers. Building up constructs from hydrogel material alone is therefore not possible using this technique.

DIW. DIW or direct–write assembly enables a wide variety of materials to be patterned in nearly arbitrary shapes and dimensions [183–185]. Colloidal based inks are patterned in both planar and 3D forms with lateral feature sizes that are at least an order of magnitude smaller than those achieved by ink-jet printing and other rapid prototyping approaches [186]. The colloidal gels are housed in individual syringes mounted on the z-axis motion stage and deposited through a cylindrical nozzle (diameter ranging from 100 µm to 1 mm) onto a moving x-y stage. The inks used in this technique must meet two important criteria. First, they must exhibit a well-controlled viscoelastic response, i.e. they flow through the deposition nozzle and then "set" immediately to facilitate shape retention of the deposited features even as it spans gaps in the underlying layer(s). Second, they must contain a high colloid volume fraction to minimize drying-induced shrinkage after assembly is complete, i.e., the particle network is able to resist compressive stresses arising from capillary tension [187].

Extruding/aspiration patterning system. The extruding/aspiration patterning system offers both extrusion, and aspiration on the basis of Bernoulli suction. The extrusion and aspiration modes are easily switchable, leading to a variety of applications in cell patterning by combining extrusion/aspiration and using a thermo-reversible hydrogel [188]. This fabrication method makes it possible to produce cell-loaded scaffolds. One of the advantages is the possibility that cell patterns can be filled into another cell matrix. A schematic illustration of an extruding/aspiration patterning system is given in Fig. 9.6. Nozzle and substrate are placed in a space that controls temperature and humidity. The nozzle is connected to the dispenser, wich regulates the pressure of the compressed air, and the valve. The system operates in three modes: extrusion, aspiration, and refilling mode. Closing or opening the valve

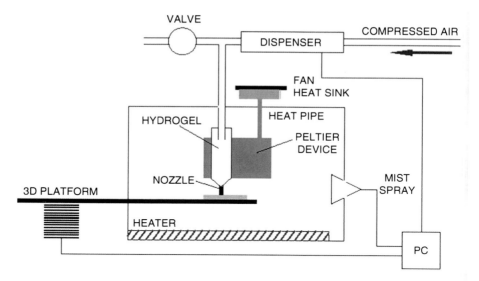

Fig. 9.6 Schematic illustration of an extruding/aspiration patterning set-up. The main difference with other nozzle-based systems is established through the incorporation of a mist spray, thus preventing desiccating of the fabricated constructs

enables switching between the modes. In a closed configuration, compressed air is supplied to the nozzle and the hydrogel is extruded onto the substrate, or refilled into the aspirated groove. By opening the valve, the compressed air is directed through the valve, and the hydrogel solution is aspirated by Bernoulli suction. To prevent desiccating (to maintain the viability of the cells), a mist spray is used to create a supersaturated atmosphere inside the box.

The most recent trend is the use of hydrogel systems and cell encapsulation strategies to fabricate gel/cell hybrid constructs [189–195]. The 3D-Bioplotter and other similar techniques can be seen as the most straightforward hydrogel/cell manufacturing method for designing complex inner and outer architectures, without the need for an additional support structure. In some cases, the developed machine contains multiple nozzles with different designs (sometimes even printer-based). In literature, "Cell Assembler" and "Bioassembly Tool" are commonly found machine names that have a similar working principle. Work of Chang et al. [196] examined the effect of nozzle pressure and size on cell survival and functionality. A quantifiable loss in cell viability was seen, which was caused by a process-induced mechanical damage to cell membrane integrity. It was suggested by the authors that cells might require a recovery period after manufacturing.

9.3.2.2 Current Limitations and Hydrogel Feasibility for Nozzle-Based Systems

An overview of the combination nozzle-based systems/hydrogel materials is summarized in Table 9.2. Lately, cell encapsulation strategies have been frequently applied to fabricate gel/cell hybrid constructs [189–193]. For each hydrogel material, it is indicated whether or not cell encapsulation experiments were performed. In addition, Table 9.5 comprises the more technical details, focussed on the disadvantages specific for hydrogels. As shown, scaffolds produced by a 3D-Bioplotter have limited resolution and mechanical strength. Material rigidity was shown to influence cell spreading and migration speed, as demonstrated by Wong et al. [197]. Cells displayed a preference for stiffer regions, and tended to migrate faster on surfaces with lower compliance. In addition, the 3D-Bioplotter technology is a time consuming technique due to the optimization of the plotting conditions for each different material. Ang et al. [179] reported 3D chitosan and chitosan-HA scaffolds using the RPBOD. Solutions of chitosan or chitosan-HA were extruded into a sodium hydroxide and ethanol medium to induce precipitation of the chitosan component. The concentration of sodium hydroxide was identified as important in controlling the adhesion between the layers. The scaffolds were then hydrated, frozen and freeze-dried. Prior to cell culturing with osteogenic cells, the scaffolds were seeded with fibrin glue. Drawbacks of this technique largely follow those of the 3D-Bioplotter and analogues. The inks used in the direct ink writing (DIW) have the disadvantage that the used hydrogel systems must satisfy two important criteria. First, they must exhibit a well-controlled viscoelastic response, i.e., they must flow through the deposition nozzle and then "set" immediately to facilitate shape retention of the deposited features even as it spans gaps in the underlying layer(s). Second, they must contain a high colloid-volume fraction to minimize drying-induced shrinkage after assembly is complete, i.e., the particle network is able to resist compressive stresses arising from capillary tension [187].

Khalil et al. [164] developed a multinozzle low-temperature deposition system with four different micro-nozzles: pneumatic microvalve, piezoelectric nozzle, solenoid valve and precision extrusion deposition (PED) nozzle. The system consisted of an air pressure supply. Multiple pneumatic valves were simultaneously operated for performing heterogeneous deposition in the development of the 3D scaffold. With this technique, multi-layered cell-hydrogel composites can be fabricated [192]. Hydrogels have also been processed with the PAM technique [170, 171]. Of the 3D rapid prototyping micro-fabrication methods available for tissue engineering, PAM has the highest lateral resolution. Recently, it has been demonstrated that the performance of this method is comparable to that of soft lithography [198]. However, capillaries with a very small diameter require careful handling to avoid any tip breakage. In addition, pressures are needed to expel the material from a small orifice. Robocasting relies on the rheology of the slurry and partial drying of the deposited layers. This implies that a pure hydrogel composition cannot be processed via this particular technique, being the most fundamental drawback of the technique. A last nozzle-based system is the extruding/aspiration patterning system. One of the advantages is that its set-up is favourable for cell encapsulation purposes and the fact that cell patterns can be filled into another cell matrix. However, the hydrogel materials require a small temperature hysteresis, so it has limited applicability.

Concerning the nozzle-based systems in rapid prototyping of hydrogels, several challenges need to be addressed. Looking at the limited range of materials, the fol-

Table 9.2 Hydrogel materials explored in nozzle-based systems

Nozzle-based systems	Hydrogel material	Cell encapsulation	References
3D Bioplotter	Gelatin based	×	[199]
3D Bioassembly tool	Gelatin/hyaluronan		
3D Cell assembler	Gelatin/alginate/fibrinogen	√	[200]
	Poly(L-lactide-co-glycolide) (PLGA) and collagen-chitosan-hydroxyapatite	×	[201]
	Hepatocytes/gelatin	√	[202]
	Methacrylated ethanolamide gelatin (GE-MA), methacrylated hyaluronic acid	√	[203]
	Gelatin/fibrinogen	√	[204]
	Gelatin/alginate	√	[205]
	Collagen precursor	×	[192]
	Gelatin/chitosan	√	[206]
	Gelatin, agar	×	[207]
	Ethylene oxide and propylene oxide based materials	×	[208]
	Poly(ethylene glycol) diacrylate (PEGDA)		
	Lutrol F127 AlaL	√	[209]
	Lutrol F127 (Poloxamer 407, BASF)	√	[43, 191]
	Pluronic F127, collagen type I	√	[191]
	Others	×	[210]
	PLLA		
	Alginate	×	[43]
	Silicone sealant Silicon SE® (product no. 057-10) based upon acetoxysilanes	×	[177]
	Polyurethane (HDI–PCL/PEG)	×	[211]

Method	Material		Ref.
RPBOD	Chitosan, chitosan-HA dissolved in acetic acid	×	[179]
DIW	Acrylamide/glycerol/water	×	[158]
	Colloidal inks contain dissolved poly (ethylene oxide)-poly(propylene oxide) (PEO; Pluronic©)	×	[212]
	PEI-coated silica microspheres suspended in water	×	[183]
	Polyelectrolyte complexes of polyanions (polyacrylic acid, PAA) and polycations (polyethyleneimine, PEI or polyallylamine hydrochloride, PAH)	×	[184]
M–LDM / (M)–LDM	Titanium diisopropoxide bisacetylacetonate (TIA)	×	[185]
	PLGA/TCP, PLGA/NaCl	×	[174]
	Gelatin/alginate/fibrinogen, gelatin/alginate/chitosan	✓	[189]
	Gelatin, sodium alginate	✓	[190]
	PLGA, PLGA/TCP; gelatin/chitosan, type I collagen	×	[165]
LDM	PLLA/TCP	×	[172]
PAM	PCL, PLLA, PLGA, PCL/PLLA	×	[167]
	PCL, PLLA, PCL/PLLA	×	[168]
	PLGA	×	[169, 198, 213, 214]
	PLLA/PCL	×	[215]
	A commercial polyurethane elastomer (Polytek 74-20, Polytek USA)	×	[216]
Robocasting	Gelatin/hepatocytes	✓	[202]
Extruding/aspiration patterning system	HA, β-TCP, HA/β-TCP, DIW, F127, corn syrup, 1-octanol	×	[181]
	Mebio gel (N-Isopropylacrylamide and polyoxyethyleneoxide)	✓	[188]

Table 9.3 Classification of inkjet printers

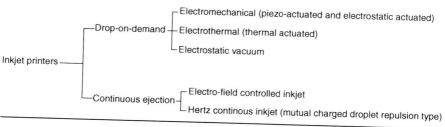

lowing topics should be addressed: optimal scaffold design, bioactivity of the scaffold as well as the issues of cell seeding and encapsulation possibilities. So, future development will need to focus on the engineering of new materials, the scaffold design and the input of cell biologists. Keeping this in mind, rapid prototyping still remains a promising technique as a methodical interface between tissue and engineering.

9.3.3 Printer-Based Systems

9.3.3.1 Working Principle and Recent Trends of Printer-Based Systems

In literature, "printing" is often used as a general term for both the construction of a scaffold or to indicate printer-based systems. To differentiate between both, we define the latter as manufacturing techniques that implement inkjet technology.

Inkjet printers can be divided in drop-on-demand or continuous ejection types. In drop-on-demand systems, electrical signals are used to control the ejection of an individual droplet. In continuous-drop systems, ink emerges continuously from a nozzle under pressure. The jet then breaks up into a train of droplets whose direction is controlled by electrical signals [217]. Both drop-on-demand and continuous-jet systems can be operated with droplets ranging in size from 15 to several hundred microns [217].

Many commercial (adapted) printers fall in the former category, and will only eject ink when receiving a demand signal from the computer. Table 9.3 classifies the existing inkjet printers (modified from Nakamura et al. [218]). Like the nozzle-based systems, building a construct occurs in an additional computer-controlled layer-by-layer sequence with deposition of material.

3DP™. Prof. Sachs from the Massachusetts Institute of Technology (MIT) introduced the 3D Printing™ technology [219]. It is an example of a solid-phase RP technology. 3D Printing can be used to fabricate parts in a wide variety of materials, including ceramic, metal, metal-ceramic composite and polymeric materials. 3D printing is the only of the solid-phase RP techniques compatible with hydrogel manufacturing. A scheme of a typical 3DP™ set-up is given in Fig. 9.7. The technique employs conventional inkjet technology. The

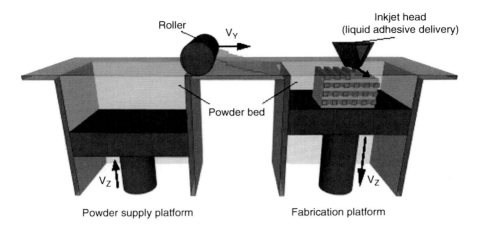

Fig. 9.7 Scheme of a typical 3DP™ set-up. A roller spreads a thin layer of polymer powder over the previously formed layer, and is subsequently solidified by the spatially controlled delivery of a liquid binder

workflow can be described in three sequential steps: (1) the powder supply system platform is lifted and the fabrication platform is lowered one layer, (2) the roller spreads the polymer powder into a thin layer (excess powder falls in an overflow vat) and (3) an inkjet print head prints a liquid binder that bonds the adjacent powder particles together. The binder can dissolve or swell the powder particles, causing bonding via the inter-diffusion of polymer chains or via infiltration of the binder into the powder [220]. Cycling steps 1–3 fabricate a 3D object. A key requirement for 3D printing is the availability of biocompatible powder-binder systems [16]. The powder utilized can be a pure powder or surface-coated powder, depending on the application of the scaffold. It is possible to use a single, one-component powder, or a mixture of different powders blended together [17]. After the finished construct is retrieved, any unbound powder is removed, resulting in a complex three-dimensional part. Basic requirements that the binder system must satisfy: (1) the binder solution must have a high binder content while still having a low viscosity so that it is capable of being deposited by the print head, (2) minimal conductivity may be required for continuous jet printing heads and (3) the binder must dry or cure rapidly so that the next layer of particles can be spread [217].

3D Printing has four steps that can limit the rate of the process: the application of the powder layers, the printing of the binder, the infiltration of the binder into the powder and the drying of the binder [217]. An important advantage of powder-based systems is the production of rougher surfaces which may enhance cell adhesion [16].

Zcorp developed a 3D printer (Z402) that uses natural polymers as well as plaster of Paris in combination with a water-based ink [220, 221]. This opens perspectives towards hydrogel manufacturing. A recent detailed review on 3DP™ concerning all process development steps, such as powder-binder selection and interaction, is given by Utela et al. [222].

Table 9.4 Hydrogel materials explored in printer-based systems.

Printer-based systems	Hydrogel material	Cell encapsulation	References
3DP™	Starch/cellulose/dextrose	×	[220, 221]
	Starch/cellulose fibre/sucrose/ maltodextrin	×	[234]
	Corn starch/gelatine/dextran	×	[221]
	Starch/polyurethanes/PEG	×	[220]
	PEO/PCL	×	[225]
	PLLGA/Pluronic® F127	×	[228]
	HA/cellulose/starch	×	[233]
Inkjet printing	Peg/Collagen/Pdl	×	[226]
	Collagen	√	[235]
	Alginate/Gelatin	√	[223]
	Fibrin	√	[227, 229]

In 3DP, control over the geometry is realized by two distinct issues: the minimum attainable feature size and the variability of part dimensions [217]. Both depend strongly on the binder droplet–powder particle interactions. Factors controlling the interaction of powder and binder include: powder material, powder surface treatment, powder size and size distribution, powder shape, powder packing density, binder material, binder viscosity, binder surface tension, droplet size, droplet velocity, temperature of the powder and binder and ambient temperature [217]. Factors that determine the final object dimensions are: local and accumulative accuracy of deposited layer thickness, accuracy of drop placement, reproducibility of the spread of the printed droplets and reproducibility of the dimensional changes that accompany binder cure. Sometimes, resolution of the machine is mentioned. Resolution in this context refers to the smallest pores and the thinnest material structures that are obtainable with the equipment [217].

Inkjet printing. This printer-based subclass comprises all liquid-phase inkjet technologies. It can vary from set-ups similar to the 3DP™ system in which the powder bed is replaced by a liquid hydrogel precursor [223], or systems that use direct inkjet writing [12, 224]. In the case of direct inkjet writing, the construct is build up by the deposited liquid itself.

9.3.3.2 Current Limitations and Hydrogel Feasibility for Printer-Based Systems

Printer-based systems can perhaps be regarded as the least hydrogel/cell suitable of the systems that allow hydrogel processing. Tables 9.4 and 9.5, summarize the hydrogel feasibility respectively limitations towards hydrogel manufacturing. Wu et al. [225] described the use of polyethylene oxide (PEO) and poly-ε-caprolactone (PCL) as matrix materials and a 20 % PCL-LPS/chloroform binder solution to create a 3D device for controlled drug release. Top and bottom layer of the tabular device was made out of slowly degrading PCL, while the interior layers were

Table 9.5 General properties summary of hydrogel compatible RP techniques

Technique	Drawbacks	Resolution	Porosity	Example[a]	References
Laser-based systems	SLA	Scaffold shrinkage due to water evaporation	30 μm	<90 %		[70, 133]
		Incomplete conversion thus post-curing essential				[236]
		Limited resolution				[138]
		Limited availability of cells, non-homogeneous cell distributions, cytotoxic photo-initiator			source: [70]	[137]
		Complex architectures with tunable micro- and macroscale features are difficult to achieve				[161]
	μ-SLA	Lower resolution than 2PP	1 μm	<90 %		[140]
		Complex architectures with tunable micro- and macroscale features are difficult to achieve				[161]
		Post-curing can be necessary			source: [1?]	
	2PP	Not feasible to produce large scaffolds	<Diffraction limit of applied light	Not specified		[133]
		Time consuming adjustment to new materials			source:[237]	[94, 220]
		Scaffold shrinkage due to water evaporation				[163]
		Diffusion driven polymerization is possible resulting in crosslinking of non-irradiated material				[163]

(continued)

Table 9.5 (continued)

Technique		Drawbacks	Resolution	Porosity	Example[a]	References
Nozzle-based systems	3D Bioplotter 3D Bioassembly tool 3D Cell assembler	Low mechanical strength Smooth surface Low accuracy Slow processing Precise control of properties of materials and medium Calibration for new material Fused horizontal pores mentioned in some cases	45–1,600 µm	<45–60 %	source: [207]	[43, 177, 209]
	RPBOD	Idem 3D Bioplotter Precise control properties of material and medium Requires freeze drying	Not specified	Not specified	source: [179]	[179]
	DIW	Low mechanical strength Smooth surface Low accuracy Slow processing Precise control of properties of materials and medium Calibration for new material	5–100 µm	< 90 %	source: [238]	[239, 240]

Method	Limitations	Size	Porosity	Image	Ref
(M)-LDM	Solvent is used Requires freeze drying	300–500 μm	75–90 %	 source: [165]	[172, 241]
PAM	Small nozzle inhibits incorporation of particle Narrow range of printable viscosities Solvent is used Highly water soluble materials cannot be used	7–500 mm	71–94 %	 source: [169]	[168]
Robocasting	Precise control of ink properties is crucial	100–150 mm	<45 %	 source: [181]	[242]
Extrusion/aspiration patterning systemw	A small thermal hysteresis of the products is required Limited applicability	141–300 mm	Not specified	 source: [188]	[242]

(continued)

Table 9.5 (continued)

Technique		Drawbacks	Resolution	Porosity	Example[a]	References
Printer-based systems	3DP Inkjet printing	Mechanical strength: post-processing often necessary	100 μm	33–60 %	 source: [220]	[177, 220, 221, 234]
		Powder entrapment				[17]
		Availability in powder form				[220, 221, 225, 228, 234]
		Pore Size: dependent on powder particle size	50–250 μm		 source: [233]	[16, 233, 243]
		Binder droplet size and accuracy of drop placement (resolution of the machine)	Droplet volume: 80–130 pL			[217, 221, 225–227, 229]
		Clogging of small binder jets				[217]
		Indirect methods needed (sacrificial moulds)	200–500 μm	50 %	 source: [230]	[230, 231]

[a]All example figures are reprinted with permission from the respective publishers as indicated in the source

composed of PEO bound by printing binder solutions. The local microstructure of the device could be controlled by either changing the binder or by changing the printing parameters (velocity). Typical powder particle size ranged from 45 to 75 μm for PCL and 75–150 μm for PEO. The binder droplets had a diameter in the order of 60–80 μm. After 20 h, significant swelling of the PEO was experimentally observed.

Landers et al. [177] studied the use of water-soluble polymers which are bonded together by means of water-based saccharide glues. Although the choice of powders and the corresponding adhesives appears to be unlimited, this technology requires post-sintering or post-curing to improve mechanical as well as environmental stability.

Lam et al. [221] developed a blend of starch-based powder containing cornstarch (50 wt.%), dextran (30 wt.%) and gelatin (20 wt.%). Distilled water was used as a suitable binder material. Cylindrical scaffolds (Ø 12.5 × 12.5 mm) were produced having either cylindrical (Ø 2.5 mm) or rectangular (2.5 × 2.5 mm) pores. Using water as the binder means that the problem of a toxic fabrication environment was eliminated and the problem of residual solvent in the construct was solved. Other advantages of using a water-based binder include the possibility to incorporate biological agents (e.g. growth factors) or living cells. Post-processing of the scaffolds was necessary to enhance the strength of the scaffolds and increase the resistance against water uptake. The scaffolds were dried at 100 °C for 1 h after printing and infiltrated with different amounts of a copolymer solution consisting of 75 % poly(L-lactide) acid and 25 % PCL in dichloromethane.

Sanjana et al. [226] report on the use of ink-jet printing to fabricate neuron-adhesive patterns such as islands and other shapes using poly(ethylene) glycol (PEG) as cell-repulsive material and a collagen/poly-D-lysine (PDL) mixture as cell-adhesive material. They worked with a positive relief: PEG used as background and cell-repulsive material was bonded covalently to the glass surface while the collagen/PDL mixture was used as the printed foreground and cell-adhesive material. They also suggest that the ink-jet printing technique could be extrapolated to building 3D structures in a layer-by-layer fashion.

Xu et al. [227] use the inkjet printing technology for the construction of three-dimensional constructs, based on fibrin gel. Fibrin was used as a printable hydrogel to build 3D neural constructs. The fibrin is formed by the enzymatic polymerization of fibrinogen by addition of thrombin and $CaCl_2$. First, a thin sheet of fibrinogen was plated and subsequently, thrombin droplets were ejected from the print cartridge onto the pre-plated fibrinogen layer. Fibrin gel formation was observed immediately after thrombin ejection. Subsequently, NT2 neurons were printed on the gelled fibrin. The whole procedure was repeated five times, resulting in a 3D neural sheet.

Koegler et al. [228] described the fabrication of 3DP scaffolds based on poly(L-lactide-co-glycolide). Surface chemistry of these scaffolds was modified by reprinting the top surface with a solution of Pluronic F127 in $CHCl_3$.

Cui et al. [229] reported on the fabrication of micron-sized fibrin channels using a drop-on-demand polymerization. A thrombin/Ca^{2+} solution together with human micro-vascular endothelial cells (HMVEC) cells was used as "bio-ink" and sprayed by the inkjet technology onto a fibrinogen substrate. They suggest that these constructs show potential in building complex 3D structures. Examples of the direct use of printer-based systems together with hydrogels are rather limited. In some cases, the use of an indirect system is mentioned.

Sachlos et al. [230] use an indirect approach to produce collagen scaffolds with predefined and reproducible complex internal morphology and macroscopic shape by developing a sacrificial mould, using 3D printing technology. This mould is then filled with a collagen dispersion and frozen. The mould is subsequently removed by chemical dissolution in ethanol and a solid collagen scaffold was produced using critical point drying.

Yeong et al. [231] also utilized a similar indirect approach to fabricate collagen scaffolds. In addition, they investigated different drying routes after removal of the sacrificial mould with ethanol. The effects of a freeze drying process after immersion of the scaffold in distilled water and critical point drying with CO_2 reflected onto dimensional shrinkage, pore size distribution and morphology in general.

Boland et al. [223] described the use of the inkjet printing technique for the construction of synthetic biodegradable scaffolds. They used a 2 % alginic acid solution, a liquid that is known to crosslink under mild conditions to form a biodegradable hydrogel scaffold. The ink cartridge was filled with 0.25 M calcium chloride ($CaCl_2$), which is known to promote the crosslinking of the individual negatively charged alginic acid chains resulting in a 3D network structure. This crosslinker was printed onto liquid alginate/gelatin solutions.

The biggest obstacles for RP technologies, thus also printer-based systems, are the restrictions set by material selection and aspects concerning the design of the scaffold's inner architecture. Thus, any future development in the RP field should be based on these biomaterial requirements, and it should concentrate on the design of new materials and optimal scaffold design [17]. The selected scaffold material must be biocompatible, compatible with the printing process, and it must be easily manufactured in the form required (powder or liquid) [232]. In the case of powder material, the particle size must be controllable. Another issue is the sterility of the manufacturing process and products and their ability to withstand sterilization processes [16]. Of course, this plays a pivotal role for all systems when embedding cells during the process.

Some limitations are caused by material trapped in small internal holes. These trapped liquid or loose powder materials may be difficult or even impossible to remove afterwards, and in some cases, these residues may even be harmful to cells and tissues. Experimental results show that the smoother the surface generated, the easier the removal of trapped material [16, 17]. Smoother surfaces are on the other hand less desirable for cell adhesion purposes.

Limitations of 3DP include the fact that the pore sizes of fabricated scaffolds are dependent on the powder size of the stock material. As such, the pore sizes available are limited to smaller pore values (<50 μm) widely distributed throughout the scaffold. More consistent pore sizes, including larger pores, can be generated by mixing porogens (of pre-determined sizes) into the powder prior to scaffold fabrication.

However, incomplete removal of the porogens is sometimes observed due to incomplete leaching. The mechanical properties and accuracy of 3DP fabricated scaffolds are other considerations that need to be addressed [233].

Despite the idea of using a water-based ink in order to eliminate a toxic fabrication environment, and thus creating an opportunity to incorporate biological agents or even living cells, toxic post-processing of the constructs is often needed to improve the mechanical properties. Suwanprateeb [234] described a double infiltration technique to increase the mechanical properties of natural polymers fabricated by three-dimensional printing using a water-based binder. The 3DP parts were porous in nature since the powder bed was only lightly packed during the process and only the surface of the powder granules was connected by a binder. Porosity typically ranged between 50 and 60 %. To enhance the performance of the 3DP parts based on a mixture of 40 wt.% starch, 15 wt.% cellulose fibre, 25 wt.% sucrose sugar and 20 wt.% maltodextrin, infiltration by some other material, was performed. The infiltration material used in this experiment was a heat-cured dental acrylate prepared by mixing triethylene glycol dimethacrylate, 2,2-bis[4(2-hydroxy-3-methacryloyloxypropyloxy)-phenyl]propane and a polyurethane dimethacrylate in a 40:40:20 wt.% proportion. Benzoylperoxide was used as initiator. After infiltration, the specimen was cured at 105 °C for 30 min. From the results, it was found that double infiltration and curing of 3DP samples increased the performance of specimens in wet conditions.

Pfister et al. [220] described the fabrication of biodegradable polyurethane scaffolds by 3D Printing. In this case, the polyurethane formation is the post-treatment step. Commercially available powder ZP11 (a powder blend of starch, short cellulose fibres and dextrose as a binder) was processed by the printing of an aqueous ink, which activates the dextrose binder. The resulting objects were very fragile and highly water-soluble. Therefore, the authors selected an additional post-treatment step involving infiltration and partial crosslinking with lysine ethylester diisocyanate. The starch polyols react with the isocyanate to form network structures.

The obtained structures exhibited much improved mechanical stability. Because of the presence of starch incorporated in the network, the structures could swell and the resulting lysine-based polyurethane networks were biodegradable upon exposure to water and body fluids.

9.4 Preserving the Mechanical Integrity of RP Processed Hydrogels

As previously mentioned, the mechanical properties of a certain biomaterial play a partial, although crucial role in its potential success as a scaffolding material. More specific, it has been generally accepted that in designing a proper scaffold biomaterial, one must strive to translate the mechanical characteristic features of the target tissue into the mechanical features of the fabricated construct. For instance, hard

Table 9.6 Classification of the most frequently applied hydrogel solidification mechanisms in RP processing

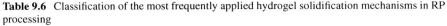
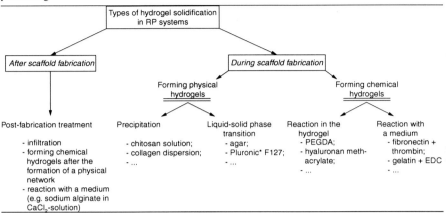

tissue regeneration requires strengths of 10–1,500 MPa, while soft tissue strengths are typically located between 0.4 and 350 MPa [244]. As a part of it, preserving the mechanical integrity of the scaffold contributes substantially to the completion of this demand. The latter appears to be of utmost importance with respect to hydrogels and will for that reason be discussed in this subsection. Moreover, it is possible to enhance the control over not only the mechanical properties, but even the biological effects and degradation kinetics by hierarchical design of scaffolds with micron to millimetre features [233]. In the case of hydrogel performance, degradation rates are controlled by hydrolysis, enzymatic reactions or simply by dissolution of the matrix (e.g. ion exchange in Ca^{2+} crosslinked alginate systems).

The process of obtaining a construct with suitable strength starts with the solidification and simultaneous shaping of the material in a certain pattern. An overview summarizing the most frequently applied and different solidification mechanisms is given in Table 9.6. Although our discussion focuses on hydrogel materials, other materials could easily be incorporated into this scheme.

The application of natural as well as synthetic hydrogels in TE and as cell embedding materials has been a review topic of several authors [11, 25, 32, 65]. The solidification or gelling mechanisms of hydrogels include inherent phase transition behaviour and crosslinking (ionic or covalent) approaches. Regarding the former mechanism, careful control over the printing temperatures can provide for some hydrogels a phase transition from solution to gel state, in particular for polymers with a lower critical solution temperature (LCST) behaviour (e.g. Pluronic® F127). However, this behaviour is reversible. The formation of an ionically crosslinked network through the use of multivalent counterions, e.g. sodium alginate and Ca^{2+} ions, provides more control over the mechanical integrity. Nevertheless, these ions could be leached out in long-term culture, or even be exchanged by other ionic molecules, compromising the control over the construct properties. Therefore, in most cases, covalent network formation is required in order to precisely enhance the mechanical

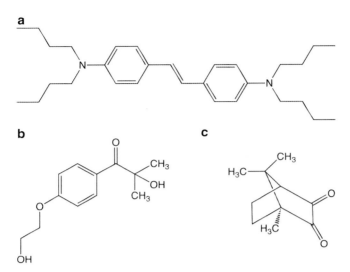

Fig. 9.8 Some typical photoinitiators used for (**a**) 2PP—an example of a strongly absorbing two photon absorbing D-π-D chromophore, (**b**) general UV curing—Irgacure® 2959, and (**c**) visible light curing—camphorquinone

stability and reproducibility of the constructs. The light-induced radical crosslinking of monomers and/or prepolymer solutions have established a quasi monopoly as a hydrogel solidification strategy in combination with RP fabrication schemes. In the case of the laser-based systems, this is even the fundament on which the respective techniques are based and those are striking examples of one-step crosslinking approaches. The chemical structures of some well-known photoinitiators are shown in Fig. 9.8. The first structure (Fig. 9.8a) is an example of a D-π-D chromophore, known for its high sensitivity in 2PP processes [245, 246]. The π-part is a conjugated backbone symmetrically substituted at the ends by electron-donating D-parts. Figure 9.8(b) represents the chemical structure of Irgacure® 2959 (I2959), a commonly applied photoinitiator thanks to its high biocompatibility. It is known to absorb light in the UV range. The last structure represents camphorquinone (CQ; Fig. 9.8c), an initiator with many dental applications and visible light working range.

Different reactive thermo-sensitive thiolated systems were developed in the past [247–254]. Thiol, vinyl or allyl ether polymerization display some advantages: mild reaction conditions, photon initiator not necessary, low or even no oxygen inhibition effects, fast process, forming of crosslinked networks with good physical and mechanical properties [255] and degradable in (mimicked) physiological conditions [256].

The so-called tandem reaction consists of two steps. The first step is the occurrence of rapid gelation kinetics. The second step involves a covalent curing based on the Michael-type addition. This technique allows to create covalently bonded hydrogels in combination with functional cells [257] under physiological conditions, with no side reactions with bioactive molecules.

When performing crosslinking in general, be it radical (light, redox, temperature induced) or non-radical (enzymatic, ionic, carbodiimide, glutaraldehyde, genipin), the toxicity of the crosslinkers should be carefully considered. This aspect is of crucial importance towards the success of a construct. For instance, comparing the biocompatibility of genipin and glutaraldehyde as a way to non-radically crosslink gelatin demonstrated the cytotoxic effect of glutaraldehyde and to a lower extent that of genipin [33, 190]. A possible aldol condensation crosslinking mechanism for genipin was proposed by Liang et al. [258], and the authors also provided evidence for its better biocompatibility compared to carbodiimide crosslinking with 1-ethyl-3-(3-dimethyl aminopropyl) carbodiimide (EDC). However, genepin has only few applications because of its high cost, and dark blue staining which could interfere with cell characterization techniques [190]. Thrombin, on the other hand, did not show any substantial adverse effects [190, 204]. Next to those non-radical crosslinking mechanisms, radical crosslinking induced through light irradiation is accepted as a common strategy. Cytotoxicity studies of several frequently used UV and visible light initiating systems have been performed [259, 260]. CQ at concentrations ≤0.01 % (w/w) and low intensity irradiation (~60 mW cm^{-2}; 470–490 nm) appeared to have less toxic effects than isopropyl thioxanthone (ITX) as visible light initiator. Another promising visible light initiator that absorbs at 512 nm is eosin Y [261, 262], although up to date no relevant cytotoxicity studies have been performed. Out of three different Irgacure® initiators, it was the I2959 initiating system (~6 mW cm^{-2}; 365 nm) that held the greatest potential at concentrations below 0.05 % (w/w). Additionally, it was also found that those cytotoxic effects varied depending on the cell types [260, 263]. Performing radical crosslinking reactions without the use of either visible or UV light can also be mediated through the use of redox systems. Duan et al. [264] reported on the negative cooperative effect of a water-soluble redox initiating system, consisting out of ammonium persulfate (APS) and N,N,N',N'-tetramethylethylenediamine (TEMED), on NIH/3T3 fibroblasts. The system was used to trigger PEG-diacrylate polymerization. Redox initiating systems have been extensively reviewed before [265].

In other cases, the formation of chemical hydrogels can only be attained after scaffold fabrication. Post-fabrication UV curing of a physically crosslinked photosensitive Pluronic F127-Ala-L construct with Irgacure® 2959 as initiator was for instance performed by Fedorovitch et al. [209] in order to acquire a chemical network. This example directly serves as an easy template into attaining multiple-step crosslinking. When working with blends of different hydrogels and combinations of physical as well as chemical network formations, multiple-step crosslinking systems of higher complexity are needed. Very recently, Xu et al. [194] demonstrated a fascinating state-of-the-art example of such an approach by blending gelatin, sodium alginate and fibrinogen with cells and stepwise crosslinking of the sodium alginate with a CaCl$_2$ solution and the fibrinogen with a thrombin solution. The gelatin component served as a necessary co-material blend in to provide sol to gel phase transition during the fabrication process [190]. Other examples can be found in the work of Xu et al. [204], Zhang et al. [199], Skardal et al. [195] and Yan et al. [205].

9.5 Fundamentals on Rapid Prototyping in Scaffold-Free Tissue Engineering

In general, the application of scaffolds in a TE approach is straightforward, but still subject to some challenges [266, 267]. These can be divided in two distinct categories: (1) complications posed by host acceptance (immunogenicity, inflammatory response, mechanical mismatch) and (2) problems related to cell cultures (cell density, multiple cell types, specific localization). It is surprisingly interesting that, during embryonic maturation, tissues and organs are formed without the need for any solid scaffolds [268]. The formation of a final pattern or structure without externally applied interventions, in other words the autonomous organization of components, is called self-assembly [269, 270]. A premise concerning the self-assembly and self-organizing capabilities of cells and tissues is worked out in the field of scaffold-free TE. This idea poses an answer to the immunogenic reactions and other unforeseen complications elicited through the use of scaffolds. One way of implementing this self-assembly concept is the use of cell sheet technology, which was demonstrated by L'Heureux and colleagues for the fabrication of vascular grafts [271, 272]. In a similar way, the group of Okano engineered long-lasting cardiac tissue based on cell sheet strategies [273–275]. Remarkably, some have already reached clinical trials [275, 276]. An alternate approach was selected by McGuigan and Sefton [33], who encapsulated HepG2 cells in cylindrical sub-millimetre gelatin modules, followed by endothelializing the surfaces. A construct with interconnected channels that enabled perfusion was generated through random self-assembly of the cell/hydrogel modules. However, the implementation of RP technologies offers another even more fascinating perspective on scaffold-free TE, and is commonly termed "bioprinting" or "organ printing".

We define organ printing as the engineering of three-dimensional living structures supported by the self-assembly and self-organizing capabilities of cells delivered through the application of RP techniques based on either laser [277–280], inkjet [218, 227, 229, 235, 281–286], or extrusion/deposition [191, 195, 287–292] technology. An emerging laser-based RP technique called biological laser printing (BioLP) stems from an improved matrix assisted pulsed laser evaporation direct write (MAPLE DW) system. The improvement is realized by incorporation of a laser absorption layer and thus eliminating the direct interaction with the biological materials. The principle is illustrated in Fig. 9.9. Prior to laser exposure, a cell suspension layer is formed on top of the absorption layer. Then, a laser beam is focused on the interface of the target, which causes a thermal and/or photomechanical ejection of the cell suspension towards the substrate [277]. Target and substrate are both able to move in the planar field.

The workflow of inkjet- or extrusion-based bioprinting can be represented by Fig. 9.10. In short, balls of bio-ink are deposited in well-defined topological patterns into bio-paper sheets. The bio-ink building blocks typically have a spherical or cylindrical shape, and consist of single or multiple cell types. Several bio-ink preparation

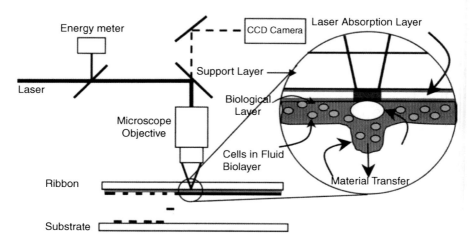

Fig. 9.9 Schematic illustration explaining the working principle of BioLP. A focussed laser beam initiates material transfer towards the substrate. Interestingly, a laser absorption layer prevents direct interaction between the laser and the biological materials. The scheme was reused from [277] with the permission from Springer

methods have been described [266, 268, 287, 291]. In a post-processing step, the construct is transferred to a bioreactor and the bio-ink spheres are fused. The bio-paper, a hydrogel, can be removed after construction if required. Bio-printers can either have a jet design or work like a mechanical extruder [101, 293]. This implies that several RP apparatuses described in the previous part can serve as a bio-printer (e.g. the Bioassembly Tool, 3D-Bioplotter™, …), if sterile conditions can be acquired. In the case of inkjet technology, individual or small cell clusters are printed. Despite the advantageous speed, versatility and cost, high cell densities are difficult to obtain and considerable cell damage is induced [266, 293]. On the other hand, extrusion-based bio-printers are more expensive but offer a more gentle approach towards cells.

In this context, hydrogels are employed as bio-paper and only provide a temporary support for the deposited bio-ink particles. In other words, the bio-paper is clearly different from scaffolds used in classical scaffold-based TE. In most cases, this bio-paper hydrogel will have a sheet-like design (e.g. Fig. 9.10). For instance, Boland et al. [294] made use of thermo-sensitive gels to generate sequential layers for cell printing. The group developed a cell printer, derived from commercially available ink-jet printers [295], that enables to place cells in positions mimicking their spatial location in an organ. The printer can put up to nine solutions of cells or polymers into a specific place by the use of specially designed software, and print two-dimensional tissue constructs. Extending this technology to three dimensions is performed by the use of thermo-reversible gels. These gels were a fluid at 20°C and a gel above 32°C and serve as bio-paper on which tissue structures can be printed. Dropping another layer of gel onto the already printed surface could generate successive layers. The thermo-sensitive gel used for the experiments was

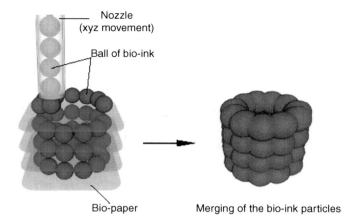

Fig. 9.10 Basic concept of bioprinting bio-ink particles into bio-paper (hydrogel) sheets. The bio-ink particles are deposited in a tubular geometry (*left*). After the deposition is finished, the construct is transferred to a bioreactor to fuse the bio-particles and further maturation made possible (*right*)

Fig. 9.11 Bioprinting tubular structures with cellular cylinders. (**a**) Designed print template (**b**) Layer-by-layer deposition of agarose (*blue*) cylinders and multicellular pig SMC cylinders (*white*). (**c**) The bio-printer outfitted with two vertically moving print heads. (**d**) The printed construct. (**e**) Engineered pig SMC tubes of distinct diameters resulted after 3 days of post-printed fusion (*left*: 2.5 mm OD; *right*: 1.5 mm OD). Pictures were reprinted from [287] with permission from Elsevier

a poly[N-isopropylacryamide-co-2-(N,N-dimethylamino)-ethyl acrylate] copolymer in a concentration of 10 wt.% polymer in cold, deionized water.

However, collagen used in a sheet-like design appeared to have integrated into the final structure, posing difficulties in its removal [291]. Depending on the target tissue design, the bio-paper can also have other geometries. For instance, agarose rods were plotted and easily removed after post-printing fusion of a

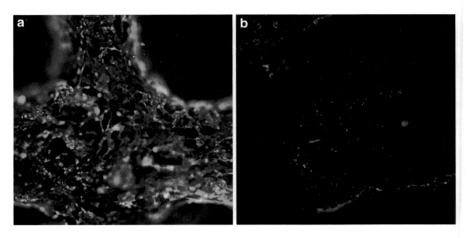

Fig. 9.12 ADC cells, cultured with EGF, in a 3D gelatin/alginate(/fibrin) scaffold to differentiate into endothelial cells and adipocytes. (**a, b**) Immunostaining with mAbs for mature endothelial cells in *green* and PI for nuclear in *red*. (**a**) ADC cells within a gelatin/alginate/fibrin construct, and (**b**) without fibrin. The pictures were reprinted from [194] with permission from Elsevier

multicellular construct in order to fabricate tubular constructs (Fig. 9.11) [287]. Agarose, being an inert and biocompatible hydrogel, is not remodelled by the cells and can easily be removed after fusion of the bio-ink. In situ crosslinkable synthetic ECM (sECM) mimetic hydrogels formed by co-crosslinking PEG-diacrylate and modified hyaluronic acid (HA)/gelatin to the corresponding thiolated dithiopropionylhydrazide (DTPH) derivatives were developed as bio-paper by Mironov et al. [293]. Other examples of hydrogels that served as bio-paper include fibrin, Matrigel™, fibrinogen, PNIPAM and polyethylene glycol tetra-acrylates [229, 235, 279, 288].

It is somewhat unclear as to whether the fabrication of cell/gel hybrid constructs, which has already been brought up in the previous part, falls under the category of bioprinting or if it can be regarded, as it is our understanding, as a bridge between pure scaffold-based and pure scaffold-free TE strategies. For instance, Skardal et al. [195] methacrylated HA and gelatin, in order to fabricate tubular photocrosslink-able constructs from partially pre-crosslinked hydrogels and cells suspended within. The importance that is correlated to the biomaterial(s) used, as is clarified in Fig. 9.12, notes the primary distinction with pure scaffold-free TE. Xu and colleagues made 3D cell/gel hybrid scaffolds from ADS cells and gelatin/alginate/fibrin hydrogels to model the metabolic syndrome (MS) in 3D [194]. Their work revealed the potential use of this 3D physiological model for drug discovery and the use of fibrin as an effective material to regulate ADC cell differentiation and self-organization into adipocytes and endothelial cells. Nevertheless the importance appointed to the biomaterial, cells are being encapsulated within the structure, and therefore this approach deflects somewhat from pure scaffold-based engineering. Maybe the future shows that this state-of-the-art TE concept and cross-over approach of the two main distinct TE premises leads to some fascinating results and research.

9.6   Future Directions

A couple of interesting emerging trends tend to bring scaffold-based TE on the next level. First of all, the implementation of gradient techniques, in the general sense, seems to provide some promising approaches. The utilization of blends layed down the foundation of working with material gradients. Via this way, a more precise mimicking of the ECM composition and mechanical properties will be possible, with spatial alterations throughout the scaffold. Of course, this can be extended towards different and multiple cell types, biomolecules, growth factors, etc., deposited in predefined patterns throughout the scaffold. Furthermore, gradients applied on the deposited scaffold pattern in itself (the deposited strand configuration) offers interesting alternatives to alter the mechanical properties of the construct. Although some authors already performed some initial experiments [165, 190], the applicability of such approaches needs to be investigated in depth.

Second, combining multiple fabrication methods to obtain a single construct appears to be useful. For instance, combining electrospinning (~nanoporosity) and bioplotting (~microporosity) for the production of a single scaffold was demonstrated by Kim et al. [296]. The combined effect of different techniques will most likely exhibit positive cooperative effects. Thus, instead of focusing on the exploitation of one single technique, it would be most fruitful to combine the positive effects of different techniques into one operation procedure.

Last but definitely the most fascinating trend, material scientists should incorporate the knowledge of engineers into the designing step of the construct. This last item is somewhat related to the first one, with a clear, distinct focus on the mechanical properties. By means of finite element modelling, predicting the mechanical properties of the construct can be handful. More importantly, adjusting the (predicted) mechanical properties of a model simply by varying the geometrical design of the material offers an interesting path to match its expected properties and the desired properties. The principles, advantages, and possible applications of this so-called Bio-CAD modelling in TE have been reviewed in 2005 by Sun et al. [223, 297]. In the last couple of years, more and more, bone-engineering scientists follow this methodology [298–300]. However, in the case of soft tissue engineering and/or tailored hydrogel scaffolds, this has not yet been intensively explored.

References

1. Wolfe RA, Roys EC, Merion RM (2010) Trends in organ donation and transplantation in the United States, 1999–2008. Am J Transplant 10(4):961–972
2. Desmet T, Schacht E, Dubruel P (2008) Rapid prototyping as an elegant production tool for polymeric tissue engineering scaffolds: a review. In: Barnes SJ, Harris LP (eds) Tissue engineering: roles, materials and applications. Nova, New York

3. Langer R, Vacanti JP (1993) Tissue engineering. Science 260(5110):920–926
4. Bonassar LJ, Vacanti CA (1998) Tissue engineering: the first decade and beyond. J Cell Biochem 30–31:297–303
5. Griffith LG, Naughton G (2002) Tissue engineering—current challenges and expanding opportunities. Science 295(5557):1009–1014
6. Chen QZ, Harding SE, Ali NN, Lyon AR, Boccaccini AR (2008) Biomaterials in cardiac tissue engineering: ten years of research survey. Mater Sci Eng R Rep 59(1–6):1–37
7. Gerlier L, Lamotte M, Wille M, Dubois D (2009) Cost-utility of autologous chon   tes implantation using chondrocelect (R) in symptomatic knee cartilage damage in i   m. Value Health 12(7):A443–A443
8. Hayashi R, Yamato M, Takayanagi H, Oie Y, Kubota A, Hori Y et al (2010) Validation   m of tissue-engineered epithelial cell sheets for corneal regenerative medicine. Tissue Eng   rt C Methods 16(4):553–560
9. Asakawa N, Shimizu T, Tsuda Y, Sekiya S, Sasagawa T, Yamato M et al (2010   vascularization of in vitro three-dimensional tissues created by cell sheet engine   g. Biomaterials 31(14):3903–3909
10. Sasagawa T, Shimizu T, Sekiya S, Haraguchi Y, Yamato M, Sawa Y et al (2010) De   of prevascularized three-dimensional cell-dense tissues using a cell sheet stacking manip   ion technology. Biomaterials 31(7):1646–1654
11. Barcili B (2007) Hydrogels for tissue engineering and delivery of tissue-inducing subst   es. J Pharm Sci 96(9):2197–2223
12. Ilkhanizadeh S, Teixeira AI, Hermanson O (2007) Inkjet printing of macromolecules on hydrogels to steer neural stem cell differentiation. Biomaterials 28(27):3936–3943
13. Chen FM, Shelton RM, Jin Y, Chapple ILC (2009) Localized delivery of growth factors for periodontal tissue regeneration: role, strategies, and perspectives. Med Res Rev 29(3):472–513
14. Ma PX (2004) Scaffolds for tissue fabrication. Mater Today 7(5):30–40
15. Butler DL, Shearn JT, Juncosa N, Dressler MR, Hunter SA (2004) Functional tissue engineering parameters toward designing repair and replacement strategies. Clin Orthop Relat Res 427:S190–S199
16. Peltola SM, Melchels FPW, Grijpma DW, Kellomaki M (2008) A review of rapid prototyping techniques for tissue engineering purposes. Ann Med 40(4):268–280
17. Yeong WY, Chua CK, Leong KF, Chandrasekaran M (2004) Rapid prototyping in tissue engineering: challenges and potential. Trends Biotechnol 22(12):643–652
18. Sachlos E, Czernuszka JT (2003) Making tissue engineering scaffolds work. Review: the application of solid freeform fabrication technology to the production of tissue engineering scaffolds. Eur Cell Mater 5:29–39, discussion 39–40
19. Martin I, Wendt D, Heberer M (2004) The role of bioreactors in tissue engineering. Trends Biotechnol 22(2):80–86
20. Stephens JS, Cooper JA, Phelan FR, Dunkers JP (2007) Perfusion flow bioreactor for 3D in situ imaging: investigating cell/biomaterials interactions. Biotechnol Bioeng 97(4):952–961
21. Awad HA, Wickham MQ, Leddy HA, Gimble JM, Guilak F (2004) Chondrogenic differentiation of adipose-derived adult stem cells in agarose, alginate, and gelatin scaffolds. Biomaterials 25(16):3211–3222
22. Benoit DSW, Schwartz MP, Durney AR, Anseth KS (2008) Small functional groups for controlled differentiation of hydrogel-encapsulated human mesenchymal stem cells. Nat Mater 7(10):816–823
23. Chenite A, Chaput C, Wang D, Combes C, Buschmann MD, Hoemann CD et al (2000) Novel injectable neutral solutions of chitosan form biodegradable gels in situ. Biomaterials 21(21):2155–2161
24. Elisseeff J, Anseth K, Sims D, McIntosh W, Randolph M, Langer R (1999) Transdermal photopolymerization for minimally invasive implantation. Proc Natl Acad Sci USA 96(6):3104–3107
25. Nicodemus GD, Bryant SJ (2008) Cell encapsulation in biodegradable hydrogels for tissue engineering applications. Tissue Eng Part B Rev 14(2):149–165

26. Williams CG, Kim TK, Taboas A, Malik A, Manson P, Elisseeff J (2003) In vitro chondro-genesis of bone marrow-derived mesenchymal stem cells in a photopolymerizing hydrogel. Tissue Eng 9(4):679–688
27. Vinatier C, Magne D, Weiss P, Trojani C, Rochet N, Carle GF et al (2005) A silanized hydroxypropyl methylcellulose hydrogel for the three-dimensional culture of chondrocytes. Biomaterials 26(33):6643–6651
28. Schmedlen KH, Masters KS, West JL (2002) Photocrosslinkable polyvinyl alcohol hydrogels that can be modified with cell adhesion peptides for use in tissue engineering. Biomaterials 23(22):4325–4332
29. Masters KS, Shah DN, Leinwand LA, Anseth KS (2005) Crosslinked hyaluronan scaffolds as a biologically active carrier for valvular interstitial cells. Biomaterials 26(15):2517–2525
30. Liu VA, Bhatia SN (2002) Three-dimensional photopatterning of hydrogels containing living cells. Biomed Microdevices 4(4):257–266
31. Tsang VL, Chen AA, Cho LM, Jadin KD, Sah RL, DeLong S et al (2007) Fabrication of 3D hepatic tissues by additive photopatterning of cellular hydrogels. FASEB J 21(3):790–801
32. Hoffman AS (2002) Hydrogels for biomedical applications. Adv Drug Deliv Rev 54(1):3–12
33. McGuigan AP, Sefton MV (2007) Modular tissue engineering: fabrication of a gelatin-based construct. J Tissue Eng Regen Med 1(2):136–145
34. van Susante JLC, Buma P, Schuman L, Homminga GN, van den Berg WB, Veth RPH (1999) Resurfacing potential of heterologous chondrocytes suspended in fibrin glue in large full-thickness defects of femoral articular cartilage: an experimental study in the goat. Biomaterials 20(13):1167–1175
35. Fussenegger M, Meinhart J, Hobling W, Kullich W, Funk S, Bernatzky G (2003) Stabilized autologous fibrin-chondrocyte constructs for cartilage repair in vivo. Ann Plast Surg 51(5):493–498
36. Matsusaki M, Yoshida H, Akashi M (2007) The construction of 3D-engineered tissues composed of cells and extracellular matrices by hydrogel template approach. Biomaterials 28(17):2729–2737
37. Yao R, Zhang RJ, Yan YN, Wang XH (2009) In vitro angiogenesis of 3D tissue engineered adipose tissue. J Bioact Compat Polym 24(1):5–24
38. Bian WN, Bursac N (2009) Engineered skeletal muscle tissue networks with controllable architecture. Biomaterials 30(7):1401–1412
39. Bryant SJ, Anseth KS (2001) The effects of scaffold thickness on tissue engineered cartilage in photocrosslinked poly(ethylene oxide) hydrogels. Biomaterials 22(6):619–626
40. Bryant SJ, Nicodemus GD, Villanueva I (2008) Designing 3D photopolymer hydrogels to regulate biomechanical cues and tissue growth for cartilage tissue engineering. Pharm Res 25(10):2379–2386
41. Endres M, Hutmacher DW, Salgado AJ, Kaps C, Ringe J, Reis RL et al (2003) Osteogenic induction of human bone marrow-derived mesenchymal progenitor cells in novel synthetic polymer-hydrogel matrices. Tissue Eng 9(4):689–702
42. Fedorovich NE, Alblas J, de Wijn JR, Hennink WE, Verbout AJ, Dhert WJA (2007) Hydrogels as extracellular matrices for skeletal tissue engineering: state-of-the-art and novel application in organ printing. Tissue Eng 13(8):1905–1925
43. Fedorovich NE, Dewijn JR, Verbout AJ, Alblas J, Dhert WJA (2008) Three-dimensional fiber deposition of cell-laden, viable, patterned constructs for bone tissue printing. Tissue Eng Part A 14(1):127–133
44. Fragonas E, Valente M, Pozzi-Mucelli M, Toffanin R, Rizzo R, Silvestri F et al (2000) Articular cartilage repair in rabbits by using suspensions of allogenic chondrocytes in alginate. Biomaterials 21(8):795–801
45. Hauselmann HJ, Fernandes RJ, Mok SS, Schmid TM, Block JA, Aydelotte MB et al (1994) Phenotypic stability of bovine articular chondrocytes after long-term culture in alginate beads. J Cell Sci 107:17–27
46. Hoemann CD, Sun J, Legare A, McKee MD, Buschmann MD (2005) Tissue engineering of cartilage using an injectable and adhesive chitosan-based cell-delivery vehicle. Osteoarthritis Cartilage 13(4):318–329

47. Hosseinkhani H, Hosseinkhani M, Tian F, Kobayashi H, Tabata Y (2006) Osteogenic differentiation of mesenchymal stem cells in self-assembled peptide-amphiphile nanofibers. Biomaterials 27(22):4079–4086

48. Iwashina T, Mochida J, Miyazaki T, Watanabe T, Iwabuchi S, Ando K et al (2006) Low-intensity pulsed ultrasound stimulates cell proliferation and proteoglycan production in rabbit intervertebral disc cells cultured in alginate. Biomaterials 27(3):354–361

49. Kisiday J, Jin M, Kurz B, Hung H, Semino C, Zhang S et al (2002) Self-assembling peptide hydrogel fosters chondrocyte extracellular matrix production and cell division: implications for cartilage tissue repair. Proc Natl Acad Sci USA 99(15):9996–10001

50. Klein TJ, Rizzi SC, Reichert JC, Georgi N, Malda J, Schuurman W et al (2009) Strategies for zonal cartilage repair using hydrogels. Macromol Biosci 9(11):1049–1058

51. Passaretti D, Silverman RP, Huang W, Kirchhoff CH, Ashiku S, Randolph MA et al (2001) Cultured chondrocytes produce injectable tissue-engineered cartilage in hydrogel polymer. Tissue Eng 7(6):805–815

52. Peretti GM, Xu JW, Bonassar LJ, Kirchhoff CH, Yaremchuk MJ, Randolph MA (2006) Review of injectable cartilage engineering using fibrin gel in mice and swine models. Tissue Eng 12(5):1151–1168

53. Roughley P, Hoemann C, DesRosiers E, Mwale F, Antoniou J, Alini M (2006) The potential of chitosan-based gels containing intervertebral disc cells for nucleus pulposus supplementation. Biomaterials 27(3):388–396

54. Stern S, Lindenhayn K, Schultz O, Perka C (2000) Cultivation of porcine cells from the nucleus pulposus in a fibrin/hyaluronic acid matrix. Acta Orthop Scand 71(5):496–502

55. Trivedi N, Keegan M, Steil GM, Hollister-Lock J, Hasenkamp WM, Colton CK et al (2001) Islets in alginate macrobeads reverse diabetes despite minimal acute insulin secretory responses. Transplantation 71(2):203–211

56. Wong M, Siegrist M, Gaschen V, Park Y, Graber W, Studer D (2002) Collagen fibrillogenesis by chondrocytes in alginate. Tissue Eng 8(6):979–987

57. Bettinger CJ, Weinberg EJ, Kulig KM, Vacanti JP, Wang YD, Borenstein JT et al (2006) Three-dimensional microfluidic tissue-engineering scaffolds using a flexible biodegradable polymer. Adv Mater 18(2):165–169

58. Anseth KS, Metters AT, Bryant SJ, Martens PJ, Elisseeff JH, Bowman CN (2002) In situ forming degradable networks and their application in tissue engineering and drug delivery. J Control Release 78(1–3):199–209

59. Bryant SJ, Cuy JL, Hauch KD, Ratner BD (2007) Photo-patterning of porous hydrogels for tissue engineering. Biomaterials 28(19):2978–2986

60. Causa F, Netti PA, Ambrosio L (2007) A multi-functional scaffold for tissue regeneration: the need to engineer a tissue analogue. Biomaterials 28(34):5093–5099

61. Chang CH, Liu HC, Lin CC, Chou CH, Lin FH (2003) Gelatin-chondroitin-hyaluronan tricopolymer scaffold for cartilage tissue engineering. Biomaterials 24(26):4853–4858

62. Dang JM, Sun DDN, Shin-Ya Y, Sieber AN, Kostuik JP, Leong KW (2006) Temperature-responsive hydroxybutyl chitosan for the culture of mesenchymal stem cells and intervertebral disk cells. Biomaterials 27(3):406–418

63. Lawson MA, Barralet JE, Wang L, Shelton RM, Triffitt JT (2004) Adhesion and growth of bone marrow stromal cells on modified alginate hydrogels. Tissue Eng 10(9–10):1480–1491

64. Lee J, Cuddihy MJ, Kotov NA (2008) Three-dimensional cell culture matrices: state of the art. Tissue Eng Part B Rev 14(1):61–86

65. Lee KY, Mooney DJ (2001) Hydrogels for tissue engineering. Chem Rev 101(7):1869–1879

66. Lutolf MR, Weber FE, Schmoekel HG, Schense JC, Kohler T, Muller R et al (2003) Repair of bone defects using synthetic mimetics of collagenous extracellular matrices. Nat Biotechnol 21(5):513–518

67. Molinaro G, Leroux JC, Damas J, Adam A (2002) Biocompatibility of thermosensitive chitosan-based hydrogels: an in vivo experimental approach to injectable biomaterials. Biomaterials 23(13):2717–2722

68. Nuttelman CR, Henry SM, Anseth KS (2002) Synthesis and characterization of photocross-linkable, degradable poly(vinyl alcohol)-based tissue engineering scaffolds. Biomaterials 23(17):3617–3626
69. Ponticiello MS, Schinagl RM, Kadiyala S, Barry FP (2000) Gelatin-based resorbable sponge as a carrier matrix for human mesenchymal stem cells in cartilage regeneration therapy. J Biomed Mater Res 52(2):246–255
70. Schuster M, Turecek C, Weigel G, Saf R, Stampfl J, Varga F et al (2009) Gelatin-based photopolymers for bone replacement materials. J Polym Sci A Polym Chem 47(24):7078–7089
71. Solchaga LA, Gao JZ, Dennis JE, Awadallah A, Lundberg M, Caplan AI et al (2002) Treatment of osteochondral defects with autologous bone marrow in a hyaluronan-based delivery vehicle. Tissue Eng 8(2):333–347
72. Thebaud NB, Pierron D, Bareille R, Le Visage C, Letourneur D, Bordenave L (2007) Human endothelial progenitor cell attachment to polysaccharide-based hydrogels: a pre-requisite for vascular tissue engineering. J Mater Sci Mater Med 18(2):339–345
73. Tsang VL, Bhatia SN (2004) Three-dimensional tissue fabrication. Adv Drug Deliv Rev 56(11):1635–1647
74. Wang L, Shelton RM, Cooper PR, Lawson M, Triffitt JT, Barralet JE (2003) Evaluation of sodium alginate for bone marrow cell tissue engineering. Biomaterials 24(20):3475–3481
75. Weinand C, Pomerantseva I, Neville CM, Gupta R, Weinberg E, Madisch I et al (2006) Hydrogel-beta-TCP scaffolds and stem cells for tissue engineering bone. Bone 38(4):555–563
76. Ahmed TAE, Dare EV, Hincke M (2008) Fibrin: a versatile scaffold for tissue engineering applications. Tissue Eng Part B Rev 14(2):199–215
77. Almany L, Seliktar D (2005) Biosynthetic hydrogel scaffolds made from fibrinogen and polyethylene glycol for 3D cell cultures. Biomaterials 26(15):2467–2477
78. Chung HJ, Park TG (2009) Self-assembled and nanostructured hydrogels for drug delivery and tissue engineering. Nano Today 4(5):429–437
79. Di Martino A, Sittinger M, Risbud MV (2005) Chitosan: a versatile biopolymer for orthopaedic tissue-engineering. Biomaterials 26(30):5983–5990
80. Farrell E, O'Brien FJ, Doyle P, Fischer J, Yannas I, Harley BA et al (2006) A collagen-glycosaminoglycan scaffold supports adult rat mesenchymal stem cell differentiation along osteogenic and chondrogenic routes. Tissue Eng 12(3):459–468
81. Tibbitt MW, Anseth KS (2009) Hydrogels as extracellular matrix mimics for 3D cell culture. Biotechnol Bioeng 103(4):655–663
82. Wang XH, Yan YN, Zhang RJ (2010) Recent trends and challenges in complex organ manufacturing. Tissue Eng Part B Rev 16(2):189–197
83. Shin H, Temenoff JS, Bowden GC, Zygourakis K, Farach-Carson MC, Yaszemski MJ et al (2005) Osteogenic differentiation of rat bone marrow stromal cells cultured on Arg-Gly-Asp modified hydrogels without dexamethasone and beta-glycerol phosphate. Biomaterials 26(17):3645–3654
84. Moller S, Weisser J, Bischoff S, Schnabelrauch M (2007) Dextran and hyaluronan methacrylate based hydrogels as matrices for soft tissue reconstruction. Biomol Eng 24(5):496–504
85. Shin H, Ruhe PQ, Mikos AG, Jansen JA (2003) In vivo bone and soft tissue response to injectable, biodegradable oligo(poly(ethylene glycol) fumarate) hydrogels. Biomaterials 24(19):3201–3211
86. Wang TW, Spector M (2009) Development of hyaluronic acid-based scaffolds for brain tissue engineering. Acta Biomater 5(7):2371–2384
87. Seo SJ, Kim IY, Choi YJ, Akaike T, Cho CS (2006) Enhanced liver functions of hepatocytes cocultured with NIH 3 T3 in the alginate/galactosylated chitosan scaffold. Biomaterials 27(8):1487–1495
88. Mann BK, Gobin AS, Tsai AT, Schmedlen RH, West JL (2001) Smooth muscle cell growth in photopolymerized hydrogels with cell adhesive and proteolytically degradable domains: synthetic ECM analogs for tissue engineering. Biomaterials 22(22):3045–3051

89. Erickson GR, Gimble JM, Franklin DM, Rice HE, Awad H, Guilak F (2002) Chondrogenic potential of adipose tissue-derived stromal cells in vitro and in vivo. Biochem Biophys Res Commun 290(2):763–769

90. Cortiella J, Nichols JE, Kojima K, Bonassar LJ, Dargon P, Roy AK et al (2006) Tissue-engineered lung: an in vivo and in vitro comparison of polyglycolic acid and pluronic F-127 hydrogel/somatic lung progenitor cell constructs to support tissue growth. Tissue Eng 12(5):1213–1225

91. LaNasa SM, Bryant SJ (2009) Influence of ECM proteins and their analogs on cells cultured on 2-D hydrogels for cardiac muscle tissue engineering. Acta Biomater 5(8):2929–2938

92. Van Vlierberghe S, Dubruel P, Lippens E, Masschaele B, Van Hoorebeke L, Cornelissen M et al (2008) Toward modulating the architecture of hydrogel scaffolds: curtains versus channels. J Mater Sci Mater Med 19(4):1459–1466

93. Fan JY, Shang Y, Yuan YJ, Yang J (2010) Preparation and characterization of chitosan/galactosylated hyaluronic acid scaffolds for primary hepatocytes culture. J Mater Sci Mater Med 21(1):319–327

94. Landers R, Pfister A, Hubner U, John H, Schmelzeisen R, Mulhaupt R (2002) Fabrication of soft tissue engineering scaffolds by means of rapid prototyping techniques. J Mater Sci 37(15):3107–3116

95. Madaghiele M, Piccinno A, Saponaro M, Maffezzoli A, Sannino A (2009) Collagen- and gelatine-based films sealing vascular prostheses: evaluation of the degree of crosslinking for optimal blood impermeability. J Mater Sci Mater Med 20(10):1979–1989

96. Noth U, Schupp K, Heymer A, Kall S, Jakob F, Schutze N et al (2005) Anterior cruciate ligament constructs fabricated from human mesenchymal stem cells in a collagen type I hydrogel. Cytotherapy 7(5):447–455

97. Schmeichel KL, Bissell MJ (2003) Modeling tissue-specific signaling and organ function in three dimensions. J Cell Sci 116(12):2377–2388

98. Dutta RC, Dutta AK (2009) Cell-interactive 3D-scaffold; advances and applications. Biotechnol Adv 27(4):334–339

99. Kirkpatrick CJ, Fuchs S, Hermanns MI, Peters K, Unger RE (2007) Cell culture models of higher complexity in tissue engineering and regenerative medicine. Biomaterials 28(34):5193–5198

100. Hutmacher DW, Cool S (2007) Concepts of scaffold-based tissue engineering-the rationale to use solid free-form fabrication techniques. J Cell Mol Med 11(4):654–669

101. Mironov V, Boland T, Trusk T, Forgacs G, Markwald RR (2003) Organ printing: computer-aided jet-based 3D tissue engineering. Trends Biotechnol 21(4):157–161

102. Roach P, Eglin D, Rohde K, Perry CC (2007) Modern biomaterials: a review-bulk properties and implications of surface modifications. J Mater Sci Mater Med 18(7):1263–1277

103. Anderson JM, Rodriguez A, Chang DT (2008) Foreign body reaction to biomaterials. Semin Immunol 20(2):86–100

104. Azuma K, Nagaoka M, Cho CS, Akaike T (2009) An artificial extracellular matrix created by hepatocyte growth factor fused to IgG-Fc. Biomaterials 31(5):802–809

105. Hubbell JA (1999) Bioactive biomaterials. Curr Opin Biotechnol 10(2):123–129

106. Kikkawa Y, Takahashi N, Matsuda Y, Miwa T, Akizuki T, Kataoka A et al (2009) The influence of synthetic peptides derived from the laminin a1 chain on hepatocyte adhesion and gene expression. Biomaterials 30:6888–6895

107. Woo JH, Kim DY, Jo SY, Kang H, Noh I (2009) Modification of the bulk properties of the porous poly(lactide-co-glycolide) scaffold by irradiation with a cyclotron ion beam with high energy for its application in tissue engineering. Biomed Mater 4(4):044101

108. Zhu YB, Gao CY, Liu XY, Shen JC (2002) Surface modification of polycaprolactone membrane via aminolysis and biomacromolecule immobilization for promoting cytocompatibility of human endothelial cells. Biomacromolecules 3(6):1312–1319

109. Zhu YB, Gao CY, Shen JC (2002) Surface modification of polycaprolactone with poly(methacrylic acid) and gelatin covalent immobilization for promoting its cytocompatibility. Biomaterials 23(24):4889–4895

110. Chu PK, Chen JY, Wang LP, Huang N (2002) Plasma-surface modification of biomaterials. Mater Sci Eng R Rep 36(5–6):143–206

111. Desmet T, Morent R, De Geyter N, Leys C, Schacht E, Dubruel P (2009) Nonthermal plasma technology as a versatile strategy for polymeric biomaterials surface modification: a review. Biomacromolecules 10(9):2351–2378

112. Tsioptsias C, Tsivintzelis I, Papadopoulou L, Pallayiotou C (2009) A novel method for producing tissue engineering scaffolds from chitin, chitin-hydroxyapatite, and cellulose. Mater Sci Eng C Biomim Supramol Syst 29(1):159–164

113. Martin L, Alonso M, Girotti A, Arias FJ, Rodriguez-Cabello JC (2009) Synthesis and characterization of macroporous thermosensitive hydrogels from recombinant elastin-like polymers. Biomacromolecules 10(11):3015–3022

114. Pathi SP, Kowalczewski C, Tadipatri R, Fischbach C (2010) A novel 3-D mineralized tumor model to study breast cancer bone metastasis. PLoS One 5(1):e8849

115. Zhu XH, Arifin DY, Khoo BH, Hua JS, Wang CH (2010) Study of cell seeding on porous poly(D, L-lactic-co-glycolic acid) sponge and growth in a Couette-Taylor bioreactor. Chem Eng Sci 65(6):2108–2117

116. Salerno A, Netti PA, Di Maio E, Iannace S (2009) Engineering of foamed structures for biomedical application. J Cell Plast 45(2):103–117

117. Gomes ME, Azevedo HS, Moreira AR, Ella V, Kellomaki M, Reis RL (2008) Starch-poly(epsilon-caprolactone) and starch-poly(lactic acid) fibre-mesh scaffolds for bone tissue engineering applications: structure, mechanical properties and degradation behaviour. J Tissue Eng Regen Med 2(5):243–252

118. Mooney DT, Mazzoni CL, Breuer C, McNamara K, Hern D, Vacanti JP et al (1996) Stabilized polyglycolic acid fibre based tubes for tissue engineering. Biomaterials 17(2):115–124

119. Liu XH, Ma PX (2009) Phase separation, pore structure, and properties of nanofibrous gelatin scaffolds. Biomaterials 30(25):4094–4103

120. Nichols MD, Scott EA, Elbert DL (2009) Factors affecting size and swelling of poly(ethylene glycol) microspheres formed in aqueous sodium sulfate solutions without surfactants. Biomaterials 30(29):5283–5291

121. Lee S, Seong SC, Lee JH, Han IK, Oh SH, Cho KJ et al (2007) Porous polymer prosthesis for meniscal regeneration. Asbm7: Adv Biomater VII 342–343:33–36

122. Wang YY, Liu L, Guo SR (2010) Characterization of biodegradable and cytocompatible nano-hydroxyapatite/polycaprolactone porous scaffolds in degradation in vitro. Polym Degrad Stab 95(2):207–213

123. Mu YH, Li YB, Wang MB, Xu FL, Zhang X, Tian ZY (2006) Novel method to fabricate porous n-HA/PVA hydrogel scaffolds. Eco-Mater Process Des VII 510–511:878–881

124. Sinha A, Guha A (2009) Biomimetic patterning of polymer hydrogels with hydroxyapatite nanoparticles. Mater Sci Eng C Biomim Supramol Syst 29(4):1330–1333

125. Mironov V, Kasyanov V, Shu XZ, Eisenberg C, Eisenberg L, Gonda S et al (2005) Fabrication of tubular tissue constructs by centrifugal casting of cells suspended in an in situ crosslinkable hyaluronan-gelatin hydrogel. Biomaterials 26(36):7628–7635

126. Pitarresi G, Palumbo FS, Calabrese R, Craparo EF, Giammona G (2008) Crosslinked hyaluronan with a protein-like polymer: novel bioresorbable films for biomedical applications. J Biomed Mater Res A 84A(2):413–424

127. Estelles JM, Vidaurre A, Duenas JMM, Cortazar IC (2008) Physical characterization of polycaprolactone scaffolds. J Mater Sci Mater Med 19(1):189–195

128. Yang SF, Leong KF, Du ZH, Chua CK (2001) The design of scaffolds for use in tissue engineering. Part 1. Traditional factors. Tissue Eng 7(6):679–689

129. Chua CK, Leong KF, Lim CS (2004) Rapid prototyping: principles and applications, 2nd edn. World Scientific, New Jersey

130. Yang SF, Leong KF, Du ZH, Chua CK (2002) The design of scaffolds for use in tissue engineering. Part II. Rapid prototyping techniques. Tissue Eng 8(1):1–11

131. Melchels FP, Feijen J, Grijpma DW (2010) A review on stereolithography and its applications in biomedical engineering. Biomaterials 31:6121–6130

132. Arcaute K, Mann BK, Wicker RB (2006) Stereolithography of three-dimensional bioactive poly(ethylene glycol) constructs with encapsulated cells. Ann Biomed Eng 34(9):1429–1441

133. Liska R, Schuster M, Infuhr R, Tureeek C, Fritscher C, Seidl B et al (2007) Photopolymers for rapid prototyping. J Coatings Technol Res 4(4):505–510
134. Lee SJ, Rhie JW, Cho DW (2008) Development of three-dimensional alginate encapsulated chondrocyte hybrid scaffold using microstereolithography. J Manuf Sci Eng Trans ASME 130(2):021007
135. Lee SJ, Kang T, Rhie JW, Cho DW (2007) Development of three-dimensional hybrid scaffold using chondrocyte-encapsulated alginate hydrogel. Sens Mater 19(8):445–451
136. Lee SJ, Kang HW, Park JK, Rhie JW, Hahn SK, Cho DW (2008) Application of microstereo-lithography in the development of three-dimensional cartilage regeneration scaffolds. Biomed Microdevices 10(2):233–241
137. Dhariwala B, Hunt E, Boland T (2004) Rapid prototyping of tissue-engineering constructs, using photopolymerizable hydrogels and stereolithography. Tissue Eng 10(9–10): 1316–1322
138. Hutmacher DW, Sittinger M, Risbud MV (2004) Scaffold-based tissue engineering: rationale for computer-aided design and solid free-form fabrication systems. Trends Biotechnol 22(7):354–362
139. Lu Y, Mapili G, Suhali G, Chen SC, Roy K (2006) A digital micro-mirror device-based system for the microfabrication of complex, spatially patterned tissue engineering scaffolds. J Biomed Mater Res A 77A(2):396–405
140. Choi JW, Wicker R, Lee SH, Choi KH, Ha CS, Chung I (2009) Fabrication of 3D biocompat-ible/biodegradable micro-scaffolds using dynamic mask projection microstereolithography. J Mater Process Technol 209(15–16):5494–5503
141. Han LH, Mapili G, Chen S, Roy K (2008) Projection microfabrication of three-dimensional scaffolds for tissue engineering. J Manuf Sci Eng Trans ASME 130(2):021005
142. Itoga K, Yamato M, Kobayashi J, Kikuchi A, Okano T (2004) Cell micropatterning using photopolymerization with a liquid crystal device commercial projector. Biomaterials 25(11):2047–2053
143. Sun C, Fang N, Wu DM, Zhang X (2005) Projection micro-stereolithography using digital micro-mirror dynamic mask. Sens Actuators A Phys 121(1):113–120
144. Gu P, Zhang X, Zeng Y, Ferguson B (2001) Quality analysis and optimization of solid ground curing process. J Manuf Syst 20(4):250–263
145. Zhang X, Zhou B, Zeng Y, Gu P (2002) Model layout optimization for solid ground curing rapid prototyping processes. Robot Comput Integr Manuf 18(1):41–51
146. Xing JF, Dong XZ, Chen WQ, Duan XM, Takeyasu N, Tanaka T et al (2007) Improving spatial resolution of two-photon microfabrication by using photoinitiator with high initiating efficiency. Appl Phys Lett 90(13):131103–131106
147. Yu T, Chiellini F, Schmaljohan D, Solaro R, Ober C (2002) Microfabrication of hydrogels for biomedical applications. In: Fedynyshyn TH (ed) Advances in resist technology and process-ing XIX, Parts 1 and 2. Spie-Int Soc Optical Engineering, Bellingham, pp 854–860
148. Mapili G, Lu Y, Chen SC, Roy K (2005) Laser-layered microfabrication of spatially patterned functionalized tissue-engineering scaffolds. J Biomed Mater Res B Appl Biomater 75B(2): 414–424
149. Arcaute K, Mann B, Wicker R (2010) Stereolithography of spatially controlled multi-mate-rial bioactive poly(ethylene glycol) scaffolds. Acta Biomater 6(3):1047–1054
150. Arcaute K, Ochoa L, Mann BK, Wicker RB (2005). Stereolithography of PEG hydrogel multi-lumen nerve regeneration conduits. ASME IMECE2005-81436 American Society of Mechanical Engineers International Mechanical Engineering Congress and Exposition, Orlando, Florida, 5–11 November 2005
151. Arcaute K, Ochoa L, Medina F, Elkins C, Mann B, Wicker R (2005) Three-dimensional PEG hydrogel construct fabrication using stereolithography. In: Fratzl P, Landis WJ, Wang R, Silver FH (eds) Structure and mechanical behavior of biological materials. Materials Research Society Sympiosium Proceedings, Warrendale, PA, pp 191–197
152. Yasar O et al (2009) A Lindenmayer system-based approach for the design of nutrient deliv-ery networks in tissue constructs. Biofabrication 1(4):045004

153. Koh WG, Revzin A, Pishko MV (2002) Poly(ethylene glycol) hydrogel microstructures encapsulating living cells. Langmuir 18(7):2459–2462
154. Khademhosseini A, Eng G, Yeh J, Fukuda J, Blumling J, Langer R et al (2006) Micromolding of photocrosslinkable hyaluronic acid for cell encapsulation and entrapment. J Biomed Mater Res A 79A(3):522–532
155. Luo Y, Shoichet MS (2004) A photolabile hydrogel for guided three-dimensional cell growth and migration. Nat Mater 3(4):249–253
156. Khademhosseini A, Langer R (2007) Microengineered hydrogels for tissue engineering. Biomaterials 28(34):5087–5092
157. Lee SJ, Rhie JW, Cho DW (2008) Development of three-dimensional alginate encapsulated chondrocyte hybrid scaffold using microstereolithography. J Manuf Sci Eng Trans ASME 130(2):021007
158. Barry RA, Shepherd RF, Hanson JN, Nuzzo RG, Wiltzius P, Lewis JA (2009) Direct-write assembly of 3D hydrogel scaffolds for guided cell growth. Adv Mater 21(23):2407–2410
159. Yuan D, Lasagni As, Shao P, Das S (2008) Rapid prototyping of microstructured hydrogels via laser direct-write and laser interference photopolymerisation. Virtual Phys Prototyping 3(4):221–229
160. Schade R, Weiss T, Berg A, Schnabelrauch M, Liefeith K (2010) Two-photon techniques in tissue engineering. Int J Artif Organs 33(4):219–227
161. Ovsianikov A, Gruene M, Pflaum M, Koch L, Maiorana F, Wilhelmi M et al (2010) Laser printing of cells into 3D scaffolds. Biofabrication 2(1):7
162. Van den Bulcke AI, Bogdanov B, De Rooze N, Schacht EH, Cornelissen M, Berghmans H (2000) Structural and rheological properties of methacrylamide modified gelatin hydrogels. Biomacromolecules 1(1):31–38
163. Ovsianikov A, Deiwick A, Van Vlierberghe S, Dubruel P, Moeller L, Draeger G et al (2011) Laser fabrication of 3D CAD scaffolds from photosensitive gelatin for applications in tissue engineering. Biomacromolecules 12(4):851–858
164. Khalil S, Nam J, Sun W (2005) Multi-nozzle deposition for construction of 3D biopolymer tissue scaffolds. Rapid Prototyping J 11(1):9–17
165. Liu L, Xiong Z, Yan YN, Zhang RJ, Wang XH, Jin L (2009) Multinozzle low-temperature deposition system for construction of gradient tissue engineering scaffolds. J Biomed Mater Res B Appl Biomater 88B(1):254–263
166. Vozzi G, Ahluwalia A (2007) Microfabrication for tissue engineering: rethinking the cells-on-a scaffold approach. J Mater Chem 17(13):1248–1254
167. Mariani M, Rosatini F, Vozzi G, Previti A, Ahluwalia A (2006) Characterization of tissue-engineered scaffolds microfabricated with PAM. Tissue Eng 12(3):547–557
168. Vozzi G, Previti A, De Rossi D, Ahluwalia A (2002) Microsyringe-based deposition of two-dimensional and three-dimensional polymer scaffolds with a well-defined geometry for application to tissue engineering. Tissue Eng 8(6):1089–1098
169. Vozzi G, Previti A, Ciaravella G, Ahluwalia A (2004) Microfabricated fractal branching networks. J Biomed Mater Res A 71A(2):326–333
170. Tirella A, Vozzi G, Ahluwalia A (2008) Biomimicry of PAM microfabricated hydrogel scaffold. Nip24/Digital Fabrication 2008: 24th international conference on digital printing technologies, technical program and proceedings 2008, pp 496–500
171. Tirella A, Orsini A, Vozzi G, Ahluwalia A (2009) A phase diagram for microfabrication of geometrically controlled hydrogel scaffolds. Biofabrication 1(4):045002
172. Xiong Z, Yan YN, Wang SG, Zhang RJ, Zhang C (2002) Fabrication of porous scaffolds for bone tissue engineering via low-temperature deposition. Scr Mater 46(11):771–776
173. Xu W, Wang XH, Yan YN, Zhang RJ (2008) Rapid protoyping of polyurethane for the creation of vascular systems. J Bioact compat polym 23:103–115
174. Liu L, Xiong Z, Zhang RJ, Jin L, Yan YN (2009) A novel osteochondral scaffold fabricated via multi-nozzle low-temperature deposition manufacturing. J Bioact Compat Polym 24:18–30

175. Khalil S, Nam J, Sun W (2004) Biopolymer deposition for freeform fabrication of tissue engineered scaffolds. Proceedings of the IEEE 30th annual northeast bioengineering conference 2004, pp 136–137
176. Zhuo X, Yongnian Y, Shenguo W, Renji Z, Chao Z (2002) Fabrication of porous scaffolds for bone tissue engineering via low-temperature deposition. Scr Mater 46(11):771–776
177. Landers R, Mulhaupt R (2000) Desktop manufacturing of complex objects, prototypes and biomedical scaffolds by means of computer-assisted design combined with computer-guided 3D plotting of polymers and reactive oligomers. Macromol Mater Eng 282(9):17–21
178. Kim GH, Son JG (2009) 3D polycarprolactone (PCL) scaffold with hierarchical structure fabricated by a piezoelectric transducer (PZT)-assisted bioplotter. Appl Phys A Mater 94(4):781–785
179. Ang TH, Sultana FSA, Hutmacher DW, Wong YS, Fuh JYH, Mo XM et al (2002) Fabrication of 3D chitosan-hydroxyapatite scaffolds using a robotic dispensing system. Mater Sci Eng C Biomim Supramol Syst 20(1–2):35–42
180. Cesarano J (1999) A review of robocasting technology. Solid Freeform Additive Fabrication 542:133–139
181. Franco J, Hunger P, Launey ME, Tomsia AP, Saiz E (2010) Direct write assembly of calcium phosphate scaffolds using a water-based hydrogel. Acta Biomater 6(1):218–228
182. Munch E, Franco J, Deville S, Hunger P, Saiz E, Tomsia AP (2008) Porous ceramic scaffolds with complex architectures. JOM 60(6):54–58
183. Smay JE, Gratson GM, Shepherd RF, Cesarano J, Lewis JA (2002) Directed colloidal assembly of 3D periodic structures. Adv Mater 14(18):1279–1283
184. Gratson GM, Xu MJ, Lewis JA (2004) Microperiodic structures—direct writing of three-dimensional webs. Nature 428(6981):386–386
185. Duoss EB, Twardowski M, Lewis JA (2007) Sol-gel inks for direct-write assembly of functional oxides. Adv Mater 19(21):3485–3489
186. van Osch THJ, Perelaer J, de Laat AWM, Schubert US (2008) Inkjet printing of narrow conductive tracks on untreated polymeric substrates. Adv Mater 20(2):343–345
187. Guo JJ, Lewis JA (1999) Aggregation effects on the compressive flow properties and drying behavior of colloidal silica suspensions. J Am Ceram Soc 82(9):2345–2358
188. Iwami K, Noda T, Ishida K, Morishima K, Nakamura M, Umeda N (2010) Bio rapid prototyping by extruding/aspirating/refilling thermoreversible hydrogel. Biofabrication 2(1):014108
189. Li SJ, Xiong Z, Wang XH, Yan YN, Liu HX, Zhang RJ (2009) Direct fabrication of a hybrid cell/hydrogel construct by a double-nozzle assembling technology. J Bioact Compat Polym 24(3):249–265
190. Li SJ, Yan YN, Xiong Z, Weng CY, Zhang RJ, Wang XH (2009) Gradient hydrogel construct based on an improved cell assembling system. J Bioact Compat Polym 24:84–99
191. Smith CM, Stone AL, Parkhill RL, Stewart RL, Simpkins MW, Kachurin AM et al (2004) Three-dimensional bioassembly tool for generating viable tissue-engineered constructs. Tissue Eng 10(9–10):1566–1576
192. Lee W, Debasitis JC, Lee VK, Lee JH, Fischer K, Edminster K et al (2009) Multi-layered culture of human skin fibroblasts and keratinocytes through three-dimensional freeform fabrication. Biomaterials 30(8):1587–1595
193. Cheng J, Lin F, Liu HX, Yan YN, Wang XH, Zhang R et al (2008) Rheological properties of cell-hydrogel composites extruding through small-diameter tips. J Manuf Sci Eng Trans ASME 130(2):021014
194. Xu M, Wang X, Yan Y, Yao R, Ge Y (2010) An cell-assembly derived physiological 3D model of the metabolic syndrome, based on adipose-derived stromal cells and a gelatin/alginate/fibrinogen matrix. Biomaterials 31(14):3868–3877
195. Skardal A, Zhang J, McCoard L, Xu X, Oottamasathien S, Prestwich GD (2010) Photocrosslinkable hyaluronan-gelatin hydrogels for two-step bioprinting. Tissue Eng Part A 16(8):2675–2685
196. Chang R, Sun W (2008) Effects of dispensing pressure and nozzle diameter on cell survival from solid freeform fabrication-based direct cell writing. Tissue Eng Part A 14(1):41–48

197. Wong JY, Velasco A, Rajagopalan P, Pham Q (2003) Directed movement of vascular smooth muscle cells on gradient-compliant hydrogels. Langmuir 19(5):1908–1913
198. Vozzi G, Flaim C, Ahluwalia A, Bhatia S (2003) Fabrication of PLGA scaffolds using soft lithography and microsyringe deposition. Biomaterials 24(14):2533–2540
199. Zhang T, Yan YN, Wang XH, Xiong Z, Lin F, Wu RD et al (2007) Three-dimensional gelatin and gelatin/hyaluronan hydrogel structures for traumatic brain injury. J Bioact Compat Polym 22(1):19–29
200. Xu MG, Wang XH, Yan YN, Yao R, Ge YK (2010) An cell-assembly derived physiological 3D model of the metabolic syndrome, based on adipose-derived stromal cells and a gelatin/alginate/fibrinogen matrix. Biomaterials 31(14):3868–3877
201. Rucker M, Laschke MW, Junker D, Carvalho C, Schramm A, Mulhaupt R et al (2006) Angiogenic and inflammatory response to biodegradable scaffolds in dorsal skinfold chambers of mice. Biomaterials 27(29):5027–5038
202. Wang XH, Yan YN, Pan YQ, Xiong Z, Liu HX, Cheng B et al (2006) Generation of three-dimensional hepatocyte/gelatin structures with rapid prototyping system. Tissue Eng 12(1):83–90
203. Skardal A, Zhang JX, McCoard L, Xu XY, Oottamasathien S, Prestwich GD (2010) Photocrosslinkable hyaluronan-gelatin hydrogels for two-step bioprinting. Tissue Eng Part A 16(8):2675–2685
204. Xu W, Wang XH, Yan YN, Zheng W, Xiong Z, Lin F et al (2007) Rapid prototyping three-dimensional cell/gelatin/fibrinogen constructs for medical regeneration. J Bioact Compat Polym 22(4):363–377
205. Yan YN, Wang XH, Xiong Z, Liu HX, Liu F, Lin F et al (2005) Direct construction of a three-dimensional structure with cells and hydrogel. J Bioact Compat Polym 20(3):259–269
206. Yan YN, Wang XH, Pan YQ, Liu HX, Cheng J, Xiong Z et al (2005) Fabrication of viable tissue-engineered constructs with 3D cell-assembly technique. Biomaterials 26(29):5864–5871
207. Landers R, Hubner U, Schmelzeisen R, Mulhaupt R (2002) Rapid prototyping of scaffolds derived from thermoreversible hydrogels and tailored for applications in tissue engineering. Biomaterials 23(23):4437–4447
208. Maher PS, Keatch RP, Donnelly K, Mackay RE, Paxton JZ (2009) Construction of 3D biological matrices using rapid prototyping technology. Rapid Prototyping J 15(3):204–210
209. Fedorovich NE, Swennen I, Girones J, Moroni L, van Blitterswijk CA, Schacht E et al (2009) Evaluation of photocrosslinked lutrol hydrogel for tissue printing applications. Biomacromolecules 10(7):1689–1696
210. Yousefi AM, Gauvin C, Sun L, DiRaddo RW, Fernandes J (2007) Design and fabrication of 3D-plotted polymeric scaffolds in functional tissue engineering. Polym Eng Sci 47(5): 608–618
211. Xu W, Wang X, Yan Y, Zhang R (2008) Rapid protyping of polyurethane for the creation of vascular systems. J Bioact Compat Polym 23:103–114
212. Xie BJ, Parkhill RL, Warren WL, Smay JE (2006) Direct writing of three-dimensional polymer scaffolds using colloidal gels. Adv Funct Mater 16(13):1685–1693
213. Kullenberg J, Rosatini F, Vozzi G, Bianchi F, Ahluwalia A, Domenici C (2008) Optimization of PAM scaffolds for neural tissue engineering: preliminary study on an SH-SY5Y cell line. Tissue Eng Part A 14(6):1017–1023
214. Mattioli-Belmonte M, Vozzi G, Kyriakidou K, Pulieri E, Lucarini G, Vinci B et al (2008) Rapid-prototyped and salt-leached PLGA scaffolds condition cell morpho-functional behavior. J Biomed Mater Res A 85A(2):466–476
215. Bianchi F, Vassalle C, Simonetti M, Vozzi G, Domenici C, Ahluwalia A (2006) Endothelial cell function on 2D and 3D micro-fabricated polymer scaffolds: applications in cardiovascular tissue engineering. J Biomater Sci Polym Ed 17(1–2):37–51
216. Tartarisco G, Gallone G, Carpi F, Vozzi G (2009) Polyurethane unimorph bender microfabricated with pressure assisted microsyringe (PAM) for biomedical applications. Mater Sci Eng C Mater Biol Appl 29(6):1835–1841
217. Sachs E, Cima M, Cornie J (1990) Three-dimensional printing: rapid tooling and prototypes directly from a CAD model. CIRP Ann Manuf Technol 39(1):201–204

218. Nakamura M, Kobayashi A, Takagi F, Watanabe A, Hiruma Y, Ohuchi K et al (2005) Biocompatible inkjet printing technique for designed seeding of individual living cells. Tissue Eng 11(11–12):1658–1666

219. Cima M, Sachs E, Fan TL, Bredt JF, Michaels SP, Khanuja S, et al (1995) Three-dimensional printing techniques. United States Patent No. 5387380

220. Pfister A, Landers R, Laib A, Hubner U, Schmelzeisen R, Mulhaupt R (2004) Biofunctional rapid prototyping for tissue-engineering applications: 3D bioplotting versus 3D printing. J Polym Sci Part A Polym Chem 42(3):624–638

221. Lam CXF, Mo XM, Teoh SH, Hutmacher DW (2002) Scaffold development using 3D printing with a starch-based polymer. Mater Sci Eng C Biomim Supramol Syst 20(1–2):49–56

222. Utela BR, Storti D, Anderson RL, Ganter M (2008) A review of process development steps for new material systems in three dimensional printing (3DP). J Manuf Process 10:96–104

223. Boland T, Tao X, Damon BJ, Manley B, Kesari P, Jalota S et al (2007) Drop-on-demand printing of cells and materials for designer tissue constructs. Mater Sci Eng C Biomim Supramol Syst 27(3):372–376

224. Sun J, Ng JH, Fuh YH, Wong YS, Loh HT, Xu Q (2009) Comparison of micro-dispensing performance between micro-valve and piezoelectric printhead. Microsyst Technol 15(9): 1437–1448

225. Wu BM, Borland SW, Giordano RA, Cima LG, Sachs EM, Cima MJ (1996) Solid free-form fabrication of drug delivery devices. J Control Release 40(1–2):77–87

226. Sanjana NE, Fuller SB (2004) A fast flexible ink-jet printing method for patterning dissociated neurons in culture. J Neurosci Methods 136(2):151–163

227. Xu T, Gregory CA, Molnar P, Cui X, Jalota S, Bhaduri SB et al (2006) Viability and electrophysiology of neural cell structures generated by the inkjet printing method. Biomaterials 27(19):3580–3588

228. Koegler WS, Griffith LG (2004) Osteoblast response to PLGA tissue engineering scaffolds with PEO modified surface chemistries and demonstration of patterned cell response. Biomaterials 25(14):2819–2830

229. Cui XF, Boland T (2009) Human microvasculature fabrication using thermal inkjet printing technology. Biomaterials 30(31):6221–6227

230. Sachlos E, Reis N, Ainsley C, Derby B, Czernuszka JT (2003) Novel collagen scaffolds with predefined internal morphology made by solid freeform fabrication. Biomaterials 24(8):1487–1497

231. Yeong WY, Chua CK, Leong KF, Chandrasekaran M, Lee MW (2007) Comparison of drying methods in the fabrication of collagen scaffold via indirect rapid prototyping. J Biomed Mater Res B Appl Biomater 82B(1):260–266

232. Sastry SV, Nyshadham JR, Fix JA (2000) Recent technological advances in oral drug delivery—a review. Pharm Sci Technolo Today 3(4):138–145

233. Leong KF, Cheah CM, Chua CK (2003) Solid freeform fabrication of three-dimensional scaffolds for engineering replacement tissues and organs. Biomaterials 24(13):2363–2378

234. Suwanprateeb J (2006) Improvement in mechanical properties of three-dimensional printing parts made from natural polymers reinforced by acrylate resin for biomedical applications: a double infiltration approach. Polym Int 55(1):57–62

235. Boland T, Mironov V, Gutowska A, Roth EA, Markwald RR (2003) Cell and organ printing 2: fusion of cell aggregates in three-dimensional gels. Anat Rec A Discov Mol Cell Evol Biol 272A(2):497–502

236. Melchels FPW, Feijen J, Grijpma DW (2010) A review on stereolithography and its applications in biomedical engineering. Biomaterials 31(24):6121–6130

237. Ovsianikov A, Deiwick A, Van Vlierberghe S, Dubruel P, Moeller L, Draeger G et al (2011) Laser fabrication of 3D CAD scaffolds from photosensitive gelatin for applications in tissue engineering. Biomaterials 12(4):851–858

238. Smay JE, Cesarano J, Lewis JA (2002) Colloidal inks for directed assembly of 3-D periodic structures. Langmuir 18(14):5429–5437

239. Simon JL, Michna S, Lewis JA, Rekow ED, Thompson VP, Smay JE et al (2007) In vivo bone response to 3D periodic hydroxyapatite scaffolds assembled by direct ink writing. J Biomed Mater Res A 83A(3):747–758

240. Geng L, Feng W, Hutmacher DW, Wong YS, Loh HT, Fuh JYH (2005) Direct writing of chitosan scaffolds using a robotic system. Rapid Prototyping J 11(2):90–97

241. Yan YN, Xiong Z, Hu YY, Wang SG, Zhang RJ, Zhang C (2003) Layered manufacturing of tissue engineering scaffolds via multi-nozzle deposition. Mater Lett 57(18):2623–2628

242. Therriault D, White SR, Lewis JA (2003) Chaotic mixing in three-dimensional microvascular networks fabricated by direct-write assembly. Nat Mater 2(4):265–271

243. Sastry SV, Nyshadham JR, Fix JA (2000) Recent technological advances in oral drug delivery—a review. Pharm Sci Technolo Today 3(4):138–145

244. Hollister SJ (2006) Porous scaffold design for tissue engineering. Nat Mater 5(7):590–590

245. Rumi M, Ehrlich JE, Heikal AA, Perry JW, Barlow S, Hu ZY et al (2000) Structure-property relationships for two-photon absorbing chromophores: bis-donor diphenylpolyene and bis(styryl)benzene derivatives. J Am Chem Soc 122(39):9500–9510

246. Kuebler SM, Braun KL, Zhou WH, Cammack JK, Yu TY, Ober CK et al (2003) Design and application of high-sensitivity two-photon initiators for three-dimensional microfabrication. J Photochem Photobiol A Chem 158(2–3):163–170

247. Cellesi F, Tirelli N, Hubbell JA (2004) Towards a fully-synthetic substitute of alginate: development of a new process using thermal gelation and chemical cross-linking. Biomaterials 25(21):5115–5124

248. Niu GG, Zhang HB, Song L, Cui XP, Cao H, Zheng YD et al (2008) Thiol/acrylate-modified PEO-PPO-PEO triblocks used as reactive and thermosensitive copolymers. Biomacromolecules 9(10):2621–2628

249. Niu GG, Du FY, Song L, Zhang HB, Yang J, Cao H et al (2009) Synthesis and characterization of reactive poloxamer 407 s for biomedical applications. J Control Release 138(1):49–56

250. Shu XZ, Liu YC, Palumbo FS, Lu Y, Prestwich GD (2004) In situ crosslinkable hyaluronan hydrogels for tissue engineering. Biomaterials 25(7–8):1339–1348

251. Du YJ, Brash JL (2003) Synthesis and characterization of thiol-terminated poly(ethylene oxide) for chemisorption to gold surface. J Appl Polym Sci 90(2):594–607

252. Brink KS, Yang PJ, Temenoff JS (2009) Degradative properties and cytocompatibility of a mixed-mode hydrogel containing oligo poly(ethylene glycol) fumarate and poly(ethylene glycol)dithiol. Acta Biomater 5(2):570–579

253. Aimetti AA, Machen AJ, Anseth KS (2009) Poly(ethylene glycol) hydrogels formed by thiolene photopolymerization for enzyme-responsive protein delivery. Biomaterials 30(30): 6048–6054

254. Govender S, Swart R (2008) Surfactant formulations for multi-functional surface modification. Colloids Surf A Physicochem Eng Asp 331(1–2):97–102

255. Lee TY, Bowman CN (2006) The effect of functionalized nanoparticles on thiol-ene polymerization kinetics. Polymer 47(17):6057–6065

256. Rydholm AE, Bowman CN, Anseth KS (2005) Degradable thiol-acrylate photopolymers: polymerization and degradation behavior of an in situ forming biomaterial. Biomaterials 26(22):4495–4506

257. Lutolf MP, Raeber GP, Zisch AH, Tirelli N, Hubbell JA (2003) Cell-responsive synthetic hydrogels. Adv Mater 15(11):888–892

258. Liang HC, Chang WH, Liang HF, Lee MH, Sung HW (2004) Crosslinking structures of gelatin hydrogels crosslinked with genipin or a water-soluble carbodiimide. J Appl Polym Sci 91(6):4017–4026

259. Bryant SJ, Nuttelman CR, Anseth KS (2000) Cytocompatibility of UV and visible light photoinitiating systems on cultured NIH/3 T3 fibroblasts in vitro. J Biomater Sci Polym Ed 11(5):439–457

260. Williams CG, Malik AN, Kim TK, Manson PN, Elisseeff JH (2005) Variable cytocompatibility of six cell lines with photoinitiators used for polymerizing hydrogels and cell encapsulation. Biomaterials 26(11):1211–1218

261. Kizilel S, Sawardecker E, Teymour F, Perez-Luna VH (2006) Sequential formation of covalently bonded hydrogel multilayers through surface initiated photopolymerization. Biomaterials 27(8):1209–1215

262. Desai PN, Yuan Q, Yang H (2010) Synthesis and characterization of photocurable polyamidoamine dendrimer hydrogels as a versatile platform for tissue engineering and drug delivery. Biomacromolecules 11(3):666–673

263. Fedorovich NE, Oudshoorn MH, van Geemen D, Hennink WE, Alblas J, Dhert WJA (2009) The effect of photopolymerization on stem cells embedded in hydrogels. Biomaterials 30(3):344–353

264. Duan SF, Zhu W, Yu L, Ding JD (2005) Negative cooperative effect of cytotoxicity of a di-component initiating system for a novel injectable tissue engineering hydrogel. Chin Sci Bull 50(11):1093–1096

265. Sarac AS (1999) Redox polymerization. Prog Polym Sci 24(8):1149–1204

266. Jakab K, Norotte C, Marga F, Murphy K, Vunjak-Novakovic G, Forgacs G (2010) Tissue engineering by self-assembly and bio-printing of living cells. Biofabrication 2(2):022001

267. Langer R (2007) Editorial: tissue engineering: perspectives, challenges, and future directions. Tissue Eng 13(1):1–2

268. Mironov V, Visconti RP, Kasyanov V, Forgacs G, Drake CJ, Markwald RR (2009) Organ printing: tissue spheroids as building blocks. Biomaterials 30(12):2164–2174

269. Whitesides GM, Boncheva M (2002) Beyond molecules: self-assembly of mesoscopic and macroscopic components. Proc Natl Acad Sci USA 99(8):4769–4774

270. Whitesides GM, Grzybowski B (2002) Self-assembly at all scales. Science 295(5564):2418–2421

271. L'heureux N, Paquet S, Labbe R, Germain L, Auger FA (1998) A completely biological tissue-engineered human blood vessel. FASEB J 12(1):47–56

272. L'heureux N, Dusserre N, Konig G, Horgan M, Kyles A, Gregory CR et al (2004) First use of a completely biological human tissue engineered blood vessel in a primate model. Circulation 110(17):508–508

273. Hannachi IE, Yamato M, Okano T (2009) Cell sheet technology and cell patterning for biofabrication. Biofabrication 1(2):022002

274. Haraguchi Y, Shimizu T, Yamato M, Kikuchi A, Okano T (2006) Electrical coupling of cardiomyocyte sheets occurs rapidly via functional gap junction formation. Biomaterials 27(27):4765–4774

275. Shimizu T, Sekine H, Isoi Y, Yamato M, Kikuchi A, Okano T (2006) Long-term survival and growth of pulsatile myocardial tissue grafts engineered by the layering of cardiomyocyte sheets. Tissue Eng 12(3):499–507

276. Nishida K, Yamato M, Hayashida Y, Watanabe K, Yamamoto K, Adachi E et al (2004) Corneal reconstruction with tissue-engineered cell sheets composed of autologous oral mucosal epithelium. N Engl J Med 351(12):1187–1196

277. Barron JA, Wu P, Ladouceur HD, Ringeisen BR (2004) Biological laser printing: a novel technique for creating heterogeneous 3-dimensional cell patterns. Biomed Microdevices 6(2):139–147

278. Guillemot F, Souquet A, Catros S, Guillotin B, Lopez J, Faucon M et al (2010) High-throughput laser printing of cells and biomaterials for tissue engineering. Acta Biomater 6(7):2494–2500

279. Guillotin B, Souquet A, Catros S, Duocastella M, Pippenger B, Bellance S et al (2010) Laser assisted bioprinting of engineered tissue with high cell density and microscale organization. Biomaterials 31(28):7250–7256

280. Wang W, Huang Y, Grujicic M, Chrisey DB (2008) Study of impact-induced mechanical effects in cell direct writing using smooth particle hydrodynamic method. J Manuf Sci Eng Trans ASME 130(2):021012
281. Boland T, Xu T, Damon B, Cui X (2006) Application of inkjet printing to tissue engineering. Biotechnol J 1(9):910–917
282. Wilson WC, Boland T (2003) Cell and organ printing 1: protein and cell printers. Anat Rec A Discov Mol Cell Evol Biol 272A(2):491–496
283. Campbell PG, Weiss LE (2007) Tissue engineering with the aid of inkjet printers. Expert Opin Biol Ther 7(8):1123–1127
284. Xu T, Jin J, Gregory C, Hickman JJ, Boland T (2005) Inkjet printing of viable mammalian cells. Biomaterials 26(1):93–99
285. Yamazoe H, Tanabe T (2009) Cell micropatterning on an albumin-based substrate using an inkjet printing technique. J Biomed Mater Res A 91A(4):1202–1209
286. Calvert P (2007) Printing cells. Science 318(5848):208–209
287. Norotte C, Marga FS, Niklason LE, Forgacs G (2009) Scaffold-free vascular tissue engineering using bioprinting. Biomaterials 30(30):5910–5917
288. Skardal A, Zhang JX, Prestwich GD (2010) Bioprinting vessel-like constructs using hyaluronan hydrogels crosslinked with tetrahedral polyethylene glycol tetracrylates. Biomaterials 31(24):6173–6181
289. Lee YB, Polio S, Lee W, Dai GH, Menon L, Carroll RS et al (2010) Bio-printing of collagen and VEGF-releasing fibrin gel scaffolds for neural stem cell culture. Exp Neurol 223(2):645–652
290. Moon S, Hasan SK, Song YS, Xu F, Keles HO, Manzur F et al (2010) Layer by layer three-dimensional tissue epitaxy by cell-laden hydrogel droplets. Tissue Eng Part C Methods 16(1):157–166
291. Jakab K, Norotte C, Damon B, Marga F, Neagu A, Besch-Williford CL et al (2008) Tissue engineering by self-assembly of cells printed into topologically defined structures. Tissue Eng Part A 14(3):413–421
292. Smith CM, Christian JJ, Warren WL, Williams SK (2007) Characterizing environmental factors that impact the viability of tissue-engineered constructs fabricated by a direct-write bioassembly tool. Tissue Eng 13(2):373–383
293. Mironov V, Prestwich G, Forgacs G (2007) Bioprinting living structures. J Mater Chem 17(20):2054–2060
294. Boland T, Mironov V, Gutowska A, Roth EA, Markwald RR (2003) Cell and organ printing 2: fusion of cell aggregates in three-dimensional gels. Anat Rec A Discov Mol Cell Evol Biol 272(2):497–502
295. Wilson WC Jr, Boland T (2003) Cell and organ printing 1: protein and cell printers. Anat Rec A Discov Mol Cell Evol Biol 272(2):491–496
296. Kim G, Son J, Park S, Kim W (2008) Hybrid process for fabricating 3D hierarchical scaffolds combining rapid prototyping and electrospinning. Macromol Rapid Commun 29(19):1577–1581
297. Sun W, Starly B, Nam J, Darling A (2005) Bio-CAD modeling and its applications in computer-aided tissue engineering. Comput Aided Des 37(11):1097–1114
298. Lian Q, Li DC, Tang YP, Zhang YR (2006) Computer modeling approach for a novel internal architecture of artificial bone. Comput Aided Des 38(5):507–514
299. Hollister SJ, Lin CY (2007) Computational design of tissue engineering scaffolds. Comput Methods Appl Mech Eng 196(31–32):2991–2998
300. Lacroix D, Planell JA, Prendergast PJ (2009) Computer-aided design and finite-element modelling of biomaterial scaffolds for bone tissue engineering. Philos Trans R Soc Lond A 367(1895):1993–2009

Index

I. Antoniac (ed.), *Biologically Responsive Biomaterials for Tissue Engineering,*
Springer Series in Biomaterials Science and Engineering 1,
DOI 10.1007/978-1-4614-4328-5, © Springer Science+Business Media New York 2013